Robert Ziegenspeck, F. H. Westerschulte

Massage Treatment - Thure Brandt - In Diseases of Women : for Practitioners

Robert Ziegenspeck, F. H. Westerschulte

Massage Treatment - Thure Brandt - In Diseases of Women : for Practitioners

ISBN/EAN: 9783337735432

Printed in Europe, USA, Canada, Australia, Japan

Cover: Foto ©berggeist007 / pixelio.de

More available books at **www.hansebooks.com**

MASSAGE TREATMENT

(THURE BRANDT.)

IN DISEASES OF WOMEN

FOR PRACTITIONERS.

.. BY ..

DR. ROB. ZIEGENSPECK,

PROFESSOR OF GYNECOLOGY AND OBSTETRICS
AT THE UNIVERSITY OF MUNICH

AUTHORIZED TRANSLATION

.. BY ..

DR. F. H. WESTERSCHULTE,

ATTENDING PHYSICIAN OF THE
NORWEGIAN LUTHERAN DEACONESS' HOSPITAL.

WITH SEVENTEEN ILLUSTRATIONS

Motto: Qui non proficit deficit.

PUBLISHED BY THE TRANSLATOR.

CHICAGO
1898.

DEDICATED

TO

MY FATHER,

SANITARY COUNCILLOR DR. H. A. WESTERSCHULTE,

GERMANY

TRANSLATOR'S PREFACE.

The medical world of to-day becomes more and more fully convinced of the fact that pelvic massage is to be classed among the most important therapeutic measures of gynecology, especially on account of the disappointment so frequently met with in the operative proceedings. The warning cry of the conservative party which looks with a skeptical eye upon operative surgery as being the panacea for woman's ills, finds a readily resounding chord in the heart of the calm and thinking physician.

On looking over the bibliography we cannot fail but notice that men who rank amongst the foremost in gynecological practice—dispassionate observers—have not only recognized and recommended pelvic massage, but have also proven its superiority to other therapeutic appliances in certain affections of the female pelvic organs. Ought not this to be a stimulus to every thinking practitioner to strive to become acquainted with a method which promises so much?

With due recognition of, aye, even enthusiasm for, the results of operative gynecology (which may with its great number of non-fatal laparotomies forcibly impress the distant observer) we nevertheless become firmly fixed in the belief that it is the mission of every therapist to cure when possible by non-sanguinous methods, before subjecting the woman to operations connected with more or less danger, and often followed by a train of disagreeable complaints. The kind reception which has been extended to mechanical therapy in gynecology is due to this gradual insight.   But the number of its opponents is by no means small, a fact which has its great advantage as it prevents every rash and unnecessary application.

It has gradually dawned upon gynecologists that mechanical difficulties must be treated according to well-known mechanical laws; and that in a great many cases the patients are not restored to health by the extirpation of the ovaries, tubes and extirpation of the uterus; and that adhesions, abdominal hernias, bodily and mental apathy, loss of physical and moral energy, so-called anovarian neurathenia, are some of the consequences of a laparotomy.

Retroflexion, fixed uterus, chronic oophoritis and perioophoritis do not justify a laparotomy, unless we have preceded it by a thorough course of massage—taking it for granted, of course, that gonorrhœa does not participate.   Laparotomy must be reserved to those cases in which the health or the life of the woman is in immediate danger.

The remedial agents heretofore used in cases of chronic para and perimetritis, connected with dislocation of the uterus, ovaries or tubes, leave a good deal to be desired in their efficiency. In pelvic massage we now possess a remedy which promises to partly or wholly fill out these therapeutical deficiencies, and a large field is offered to the practitioner for its employment.

In translating this work the German text has at all times been closely followed, sometimes even at the risk of the English idiom.

684 W. North Ave., Chicago, Ill.

PREFACE TO THE AUTHORIZED ENGLISH TRANSLATION.

My colleague, Mr. Westerschulte, has undertaken the praiseworthy task of translating my book into English, in order that both physicians and patients on the other side of the Atlantic ocean may become better acquainted with a therapeutical method, calculated to save many suffering women from permanent illness. I gladly comply with the wish of the translator to preface this work with a few commendatory words, adding a portrait of Major Thure Brandt, whose therapeutic invention I have tried to establish on a scientific basis and to elucidate in this book in the simplest possible manner. The dear features of my "old friend," as he was wont to sign himself in his frequent epistolary correspondence with me, impart life to the cold letters, thus creating the impression in the mind of the reader as if he himself were speaking to us. Mr. Brandt, an ever-active, restless man, died in August, 1895, at the age of seventy-three years, in the midst of his professional and literary pursuits. He was ever desirous both of learning and teaching. Up to his death he kept up an instructive scientific correspondence with the author. For more than a year, now, he has been reposing in his last resting place. To give this method the largest possible publicity for the benefit of suffering women, is immeasurably better than the erection of a monument, and in all likelihood more according to the wishes of this remarkable man.

In this sense, therefore, I warmly press the hand of the translator, Dr. Westerschulte, across the Atlantic ocean.

My method differs essentially from that laid down in numerous other works on this subject. Proceeding from the simple to
the complicated, not only the method per se, but also the diseases indicated are described; whereas, the other works at most
give only extracts of Brandt's own book about his method, or
simply contain more or less practical suggestions for the modification of a part of the method.

From Brandt's book, entitled '' Hygienic Gymnastics of the
Female Sexual Organs,'' my work differs not only in so far as it
discusses causes, complaints and state of diseases in which the
method is indicated, but especially in the fact that it describes
the method in the simplest possible manner, omitting everything
superfluous or irrelevant, but mentioning and appreciating at the
same time the competitive methods heretofore generally used in
medical and gynecological practice.

While Brandt, in describing his method, lays particular
stress upon hygienic gymnastics as general treatment against the
reflex symptoms, aside from the local gynecological treatment
discovered by him, the author, as physician, describes, according
to his own pathologic-anatomical researches, corroborated by
clinical experience, the conditions as found, the symptoms dis-
appearing by themselves after the removal of the latter and after
the restoration of the normal anatomy.   Accordingly the author
stretches the parametric bands, in order to liberate the vessels
and nerves from contracting cicatricial tissue, thus restoring the
normal function; whereas Brandt thinks he is applying hygienic
gymnastics to the ligaments of the uterus (resistance movements).
The author being aware that the only disadvantage of his method
consists in its tedious details, has striven in every instance to
omit everything superfluous and irrelevant, in order to avoid

these inconveniencies as much as possible. In such cases only in which the method cures more surely, quickly and pleasantly, will it assert itself and stand the competition of more convenient methods. In this regard, too, I have pointed out the limits of its indication in this book.

In view of the fact that chronic parametritis, with or without retroflexion, can be more easily distributed, and that the chances of recovery—even of the displacements—are more favorable if the treatment is begun soon after its genesis, considering, moreover, that every peritonitic adhesion or pseudomembrane is, in its incipient stage, of gelatinous softness and therefore easily removable, but later on becomes so solid that it can hardly be torn, it can readily be understood that an attempt for its removal ought to be made at the earliest possible convenience.

Not until every disease of this kind, three or at most six months after its formation, has been dispelled by the physician's skillful hand, will this method celebrate its greatest triumphs. However, the complaints being often only slight shortly after the genesis of the disease, it becomes the duty of the practitioner who has to treat puerperal diseases, gonorrhœa, and other ailments, to call the attention of the patient to the necessity of her submitting to an examination at a time when we are justified in assuming that the germs of the disease have become " innocuous," in order to establish the presence of any sequelæ. If this course is followed out, any so-called pathological anteflexions, retroflexions, or cystocolpoceles may easily be cured by massage and stretching. It is for this reason that the method, or at least a knowledge of its indications, devolves upon the practitioner. He should instruct his patients as to the possibility of a cure if the disease is discovered in its early development, or a well-nigh impossible cure if it is not seen until its subsequent stages. Then will be ushered in a time when the happiness of the family is no longer disturbed by a sick woman, when the husband has his wife and the children have their mother.

That the method is tedious and laborious is no disadvantage then, we being amply repaid for our labor.   Let us only bear in our mind that we have to deal with women in the prime of their life, and that the happiness of whole families is involved.

Perhaps—I repeat it for the third time—the excessively large number of physicians, so universally complained of nowadays, may be profitably and successfully employed in this manner.

DR. ROBERT ZIEGENSPECK,
Privatdocent at the Royal Ludwig Maximilian University, Munich.

INTRODUCTION.

Some of the readers of this work who have become more or less familiar with massage treatment in gynecology, will have noticed that the general interest aroused some years ago by Thure Brandt's therapeutic method is waning or at least flagging. A great number of publications have appeared bearing on this subject, the object of which was, at first, merely to give a description of the method, and, subsequently, to improve, or, more accurately speaking, to simplify it.

How did it happen, our kind reader may ask, that so many authors found it necessary to describe the method more fully, inasmuch as Brandt himself had already published an entire book on the subject? Nothing else was needed but a translation of the same into German. The causal connection was as follows:

When Brandt returned from Jena, where he had demonstrated his method and treated, under the latter's supervision, those well-known fifteen cases reported by Profanter, Dr. Resch, formerly assistant to Saenger and Schultze, determined to learn Brandt's method from Brandt himself, accompanied the latter to Stockholm, Sweden. While in Stockholm, Resch attended not only Brandt's clinic, but also the Royal Central Institute for Hygienic Gymnastics. This institute has the same import for Sweden as the Central Gymnasium for the German army, the army officers in Sweden being sent from the special garrisons to this institute in order to acquire a knowledge of gymnastics, which in the Swedish army take the place of calisthenics. Incidentally, many Swedish officers thus voluntarily acquire a knowledge of that branch of hygienic gymnastics which has for its purpose the cure of certain diseases.

While residing in Stockholm, Dr. Resch provided for a translation of Brandt's book on "Gymnastics" into the German language, a French edition, from the original Swedish text, having already been published. I use the word "provided" intentionally, for the real work of translation was done by a teacher sent for from Germany. Resch's share in the so-called translation consisted merely in omitting a large part of Brandt's therapy, regarded as useless, or, at least, unessential, and in altering certain gynecological terms. These verbal alterations, however, did not always represent exactly what Brandt had intended to convey by them. The old major felt very much chagrined over this arbitrary proceeding, and called the attention of everyone coming to him for instruction in massage to his (Brandt's) original manuscript. The latter now became the source of those numerous publications mentioned above. Nearly all who examined the manuscript detected in it something not yet published, but worth mentioning, and did not hesitate to make their discoveries known. This condition of things continued until Professor Schauta brought out a second edition, without essential abridgment and with the introduction of suitable gynecological terms, together with a great many excellent illustrations. To him who wishes to enter more deeply into the subject, and, above all, to him who wants to determine the full limit of its application, this book is indispensable. In publishing this work I do not intend it to supersede Brandt's book, but, rather, to extend the knowledge of his method by means of a compact, lucid and yet strictly scientific treatise as widely as possible among practitioners, and to direct their attention to a certain class of cases in which this treatment stands unexcelled.

Unlike any other, this mode of treatment, as employed by
the medical practitioner, is apt to prove an inexhaustible source
of blessings, whereas it is unfit for the coryphæi of science—for
the simple reason that it requires too much time and labor.

If the appearance of the second edition brought out by Professor Schauta was a sufficient reason for the fact that special
literature had become almost silent on this subject, the fact last
mentioned above is a second, and, as it seems to me, the chief
reason why our very first authorities ignore this method or even
assume a chilly and declining attitude. From this, however,
one need not necessarily infer that they are prejudiced against
it, or unwilling to grant it full justice. The explanation of this
mental disposition is rather to be looked for in the peculiar
division of labor into special branches. The professional gynecologist of to-day is a surgeon, preferably an abdominal surgeon.
He is a master of that branch of surgery which requires the
most careful antiseptic and aseptic measures. Although he may
be profoundly conscious of his second mission, i. e., to teach his
pupils, who will later become general practitioners, the simplest
and best methods for the care of female troubles, yet circumstances are such that his skill as an operator and the safety of
his antiseptic and aseptic devices constitute the first claim upon
his attention. Coming mostly from a circle of patients varying
in extent according to the size of the city or university, the
operative material collects about him, for the greater part
patients wishing to be or obliged to be operated upon. The
result is that he is gradually becoming weaned from bloodless
methods. Thus it happens that at meetings of gynecologists
specimens of diseased ovaries and tubes are exhibited with tri-
umphant air, as if intended to prove the excellent results of
radical operations (i. e., recovery from grave operative manipulations), whereas in the same time, by means of massage, etc.,

the disease with all its troubles might have been removed and the organs preserved. Moreover, after the operation the symptoms of the patient and even danger of resultant abdominal hernia persist for a long time. Massage, however, would have cost the operator a great deal more time and he would have achieved less renown. Laparotomy is still considered a master performance, and many operations are undertaken for the glory of the operator rather than for the benefit of the patient.

Having thus demonstrated, I think, why "silence reigns again" over Brandt's treatment, and why the very chieftains in medical science fail to recommend it, I must not forget to acknowledge that a great many practicing physicians and even lesser gynecologists, i. e., non-laparotomists, employ the method with the best of results. In Munich, especially, a considerable number of physicians treat retroflexions and the so-called pathological anteflexions according to Brandt's method with the greatest success. Among them I am pleased to mention Drs. Rudolph Mueller, Stieler, Doldy, Faust, Stern, Fuld, Cornet, Mottes, Raimund Maier, Heintz, Von Zetschwitz, Schmidt and others. I am teaching the method in my lectures at the university. How many of my auditors employ it later on in their practice, it is hard to tell; but I am sure that whoever may have employed it, and in appropriate cases compared it with other methods, will continue to use it.

At the time the author wrote his paper for "Volkmann's Vortraege" in 1888, he had treated sixty cases. By the time the book appeared this number had increased to one hundred. Since that period about a thousand cases have been added, treated partly in private practice and partly with his assistants in the clinic. It would seem that the number of patients and the length of the period of observation were sufficient to cool any too great enthusiasm which a new method is apt to arouse, and to settle any doubts as to the permanence of the results of the treatment. My conscience would revolt were I to fail to recommend massage in even a single instance in which I have employed it in the past. No method cures so quickly, none has so few relapses. The writer of this book numbers himself among those authors who have striven to simplify the method, for, as previously stated, its excessive detail, demanding a great outlay of time and labor, constitutes its only disadvantage. Only in cases in which the treatment, as I remarked before, cures safely, quickly and pleasantly in a higher degree than other gynecological methods, would it be able to compete with the latter; but it would certainly fail in cases in which the results of other methods are simply *equal* to those of massage by virtue of their being more convenient. Since that time I have, from this standpoint, more carefully examined my rather extensive material for observation, and now, after more than five years, my views regarding the conditions demanding massage treatment and scientific explanation of its effects remain absolutely unchanged. Excepting for the fact that my judgment has in certain matters become more accurate, I need not retract one word from what I said in former discussions. Though unable, therefore, to offer anything essentially new in the following pages, yet friendly hints, calling my attention to the fact that the method is being less discussed than formerly, induce me to substantiate the correctness of the statements heretofore made by a larger number of cases. As to the permanence of the results, too, my judgment is entitled to some consideration. I was the first in Germany to treat cases according to Brandt's own directions. As early as in the winter of 1886–87, when Brandt was still in Jena, I treated at my native home in Thuringia, where I was spending my vacation, a number of patients according to Brandt's teachings. With scarcely any exception they have retained their health up to the present time. And to-day yet I am of the opinion that the essential element in Brandt's treatment consists in massaging and stretching of chronic contracted inflammatory processes, although Brandt still adheres to his original view, that the benefit is due to the effect of hygienic gymnastics on the uterine ligaments. This difference of opinion, however, will not engender any discord between Brandt and myself, Brandt, as heretofore, taking comfort in the fact that ultimately it is immaterial to him as to how the method acts, provided only its therapeutic efficiency be acknowledged.

HYGIENIC GYMNASTICS AS GENERAL TREAT-
MENT.

I still adhere to my previous opinion, that hygienic gymnastics bear the same relation to local treatment that general
hydrotherapeutic or electric treatment bears to our local gynecological therapy, and that hygienic gymnastics may be substi-
tuted, as experience has shown, by machine gymnastics or by
balneotherapy. The so-called reflex symptoms disappear more
rapidly if local treatment is combined with hygienic gymnastics.
Machine gymnastics and hydrotherapeutic measures accomplish
the same, the former at times even in a higher degree.

Those who are not willing to accept my opinion, although
based upon experience, but prefer following their own judgment,
will certainly be obliged to proceed strictly according to Brandt's
method. Perhaps my paper in ''Volkman's Vortraege,'' old
series, No. 353–354, will be of some benefit to them. For the
first year and a half I strictly followed Brandt's method, and I
must confess that, with the aid of hygienic gymnastics, the
so-called reflex symptoms frequently disappeared before the etiological factor of the trouble had been removed, whereas without
them the reflex symptoms often persisted and disappeared
only gradually. I employed the treatment also in such a manner as to use hygienic gymnastics every alternate day or every
alternate week, and asking the patients whether they noticed
any difference. I found that hygienic gymnastics could be dispensed with in a great many cases, especially in women who
were still moving about and attending to their daily routine of

domestic duties. To such women a few dozen gymnastic movements mean as much as the addition of a few drops of tokay
as a remedial agent to a professional wine drinker who imbibes
one or two bottles of wine daily. Hygienic gymnastics are,
however, indispensable to those women who, compelled by
their overconscientious physicians to remain in bed for months
or even years and kept therein to their own detriment by over-
zealous relatives, have lost all confidence in their muscular and
nervous apparatus. Such patients, when suffering from hystero-
cataleptic attacks, often manifest such strength that more than
two persons are required to keep them in bed, whereas they are
hardly able to walk two steps. In others, as a result of com-
pulsory inactivity real muscular atrophy has set in. Aboulia
(loss of will power), in the broadest sense of the word, is at
the bottom of the evil in a majority of cases—a mental affection so common among patients suffering from abdominal diseases. The usual excuse that "I cannot," being the principal
feature of hysteria, can, however, be overcome by the will of
another person or by stern necessity. When the experienced
physician sees that one patient afflicted with chronic parametritis,
although complaining of downward pressure, vesical tenesmus,
pain in the side, constipation, and only occasional pain in
the thigh interfering with walking, is nevertheless attending to
her household duties, while another with just the same local
affections remains in bed for months and years, opposing every
invitation to take exercise with the stereotyped phrase, "I cannot," imagining herself to be dreadfully sick and fearing an
aggravation of her trouble by even the slightest bodily exercise,
he may well regard this as an admonition not to be too liberal
in prescribing confinement to bed for such patients.

In those cases, however, in which it is too late to prevent
confinement to bed, we can effectually employ manual gymnas-

tics, thereby putting the patient again on her feet. At the beginning a determination on the part of the patient is necessary before the execution of any single movement, every one of them meaning stimulation of will power. Machine gymnastics and balneotherapy can be employed only after local treatment in order (1) to remove the atrophy completely, (2) to familiarize the patient again with the co-ordinate movements, and (3) to dull the general nervous and muscular apparatus against the abnormal impulses of motion.

It is supposed to be known that Kinesiotherapy (movement cure) consists of: (1) Active movements, executed by the patient himself or by the patient aided by an assistant; (2) passive movements, performed by the assistant on the patient; (3) acts of resistance to movements, whether executed by the assistant against the patient or by the patient against the assistant. In addition to these movements we have percussion, chopping, stroking, pressure and laying on of hands. As a special subclass of the passive and resistance movements are to be mentioned "respiratory movements." Nearly every movement of artificial respiration is here represented. As a subspecies of pressure we may put down vibration pressure (Zitterdruck). While the former are destined to play a prominent part in the treatment of cardiac affections, they are without direct importance in the treatment of abdominal diseases, except when a disease of the respiratory or circulatory apparatus is in direct relationship to ailments of the sexual organs. They may be found mentioned in nearly every prescription* ordering movements for drawing the blood away from the pelvic organs. If there are really movements able to divert the blood from the pelvis, an action

* Brandt himself carefully avoids this term. It is used only in the Central Gymnastic Institute, in Stockholm, on the card upon which number, kind and succession of each single movement are recorded.

2

most desirable in all pelvic inflammations, the respiratory movements ought to rank first among them.   The latter, by promot-
ing the action of the muscle, not only stimulate the circulation
(by increasing the volume of blood in the arteries and diminishing it in the veins), but also promote deeper respiration, the
vascular system of the lungs, sponge-like, absorbing the blood
from the right heart during inspiration and emptying it out into
the left heart during expiration, besides increasing the exchange
of gases in the lungs and thereby furthering metabolism.   As an
example of respiratory movements let us take "relaxed posture,
chest expansion" (Schlaffsitzend, Brusthebung).

POSITION.—The patient is sitting in a relaxed position upon
a chair, her arms hanging down and her head bent backwards.
The physician is standing behind the patient upon a chair, one
foot turned inward close to the buttock of the patient and knee
pressed between the shoulder blades.   The hands encircle the
shoulder from the front and above.

MOVEMENT.—The patient taking a deep breath, her shoulders and arms are pulled vigorously upwards and backwards, so
that the thorax is raised, and, by being pressed against the
knee of the physician, pressed forward.   During expiration the
involved parts are to be released, and the movement is to be
repeated four or five times.   The movement is a well-nigh exact
imitation of Sylvester's artificial respiration, and its effect,
"deepening of breathing," is the same.

As an example of resistance movements we may select
stretch-support-stride-sitting (streckneigspaltsitzend) and double
arm-bending (doppelte Armbeugung).*

POSITION.—The patient is sitting straight upon a chair, her
arms raised and thighs spread out.   The attendant is standing
in front of the patient upon a chair, grasping with both hands

---

*Brandt, Treatment of Women's Diseases.

the forearms of the patient and the patient in her turn grasping the forearms of the attendant.

MOVEMENT.—The patient, flexing her elbow in an outward direction, draws on her arms under resistance of the attendant. Thereupon the attendant draws her arms upwards and straightens them under resistance of the patient.

EFFECT.—The blood is strongly diverted from the pelvis and from the head towards the back.

Everyone of these movements may also be classed under passive movements. Let us assume, for instance, a case of central paralysis of the arm. We would first use passive movements, so that, while the nerves are inactive, the muscles may not become atrophied and the joints remain in exercise. By and by other ganglia for the respective muscular function are formed centrally, and soon the patient is able to move his arm actively. A little resistance even may be used afterwards, which is gradually increased, until the arm is nearly or entirely cured. We employ a similar treatment in bone, muscle and joint injuries, only with the difference that here massage, i. e., the mechanical support of the circulation, is of greater moment, while in the former stimulation of peripheral nerves by compression, percussion, etc., is regarded as the prime factor.

As especially effective in increasing the circulation must be mentioned the "revolving movements" (Rollbewegungen). As an example we may take: Half-lying, foot-revolving (Fussdrehung).

POSITION.—The patient lies in a semi-recumbent position with her back upon a bench. The attendant sits at her feet and the lower leg of the patient rests upon his knee. With one hand he encircles the tip of her toes, and with the other hand her heel.

MOVEMENT.—The physician describes circles with the tip of

the patient's foot, first about ten from left to right, then as many from right to left, without resistance on the part of the patient. The same is done with the other foot.

EFFECT.—In case of amenorrhœa the blood is directed towards the pelvic organs, in case of cold feet warmth is restored. The author had this method tried once on himself by Brandt while suffering from extremely cold feet during winter time, and became thoroughly convinced of its powerful influence upon the circulation. These movements are also called "pumping movements." A negative pressure occurs in the veins of that side which is stretched during the operation and the blood is aspirated; while on the flexed side the blood is conveyed beyond the valves of the veins by compression, thus producing an effect similar to that of a pump.

Percussion, chopping, compression, stroking, and even the simple laying on of hands are stimulants which have to be selected and modified according to the thickness and other conditions of the tissues upon which they are intended to have an effect. They all produce active hyperæmia and in this way improve the nutrition of the respective regions. Furthermore, they may be applied to relieve pain, as compression of the superciliary nerve in "tic douloureux." For the most part they produce pain momentarily and serve to increase innervation and nutrition.

Although we have no difficulty in understanding that exactly the same stimulation can be exercised by mechanical devices as by electric or thermic means (electrotherapy and hydrotherapy), yet it is not quite clear to us from the start how stroking and laying-on of hands should have any effect. We must know, then, that these manipulations are performed with the aid of psychic forces.

The patient keeps his eyes closed, and is told to follow every

one of the movements with the utmost attention. She is asked
repeatedly whether such and such part of the body has not yet
become warmed, or if the pain has not ceased. Brandt, while
laying on his hands, occupies a kneeling position, fervently
imploring God for help. (In Stockholm he is known as a pious
man, in the best sense of the word.) In this way he has
frequently cured hysterocataleptic attacks—however, mostly in
credulous individuals. Brandt considers the effect produced by
the imposition of the hands as a magnetic phenomenon, as
Messmer did, and denies that it might be explained more plausi-
bly as suggestion in a waking or at most slightly hypnotic condition.

In regard to compression, it must be added that in the form
of vibration pressure it is less painful and yet more effective
than steady presure. Thus, in case of floating kidney, upward
pressure of the kidney is exercised as vibratory pressure, and
the kidney can be pushed with slight pain farther beneath the
liver, into the hypochondriac region.

The compression of veins has a special purpose, as, for
instance, the compression of the jugular vein in headaches. It
is supposed that, through some irritation, the afflux of arterial
blood is increased, but the venous reflux decreased; hence cere-
bral congestion and headache. If we compress now the jugular
vein temporarily, we mechanically dilate by retrogressive congestion the venous system as far as the capillaries, and thus
facilitate the venous reflux; the congestion is relieved, and with
it the headache.

It must be added here, that by the chopping and percussion
movements, applied by Brandt almost exclusively to the back
along the spinal column, quite a different effect is supposed to
be produced according as they are applied energetically or only
slightly, in strained or in easy position. In an easy position,

for instance, "Stuetzneiggegenstehend," that is, slightly bent
forward, with the hands resting at about shoulder height upon
the opposite wall, slight choppings on the back and percussion
transversely across the sacral region are said to induce the central nervous system to exercise a tonic influence upon the relaxed ligaments of the uterus, while in a strained position, that
is, with such a concave back that the patient is obliged to support herself with her raised arms against the wall, these movements are said if energetically applied, to produce an afflux of
blood toward the pelvis, which might even produce abortion.

If we ask what proofs there are for such effects, the answer
is that the results have been satisfactory in so many cases.
However, not only one movement, but half a dozen and more,
are made use of in the treatment.

"To supply innervation and nutrition," "to vitalize," "to
conduct," "to divert"—these are the four watchwords of the
Swedish hygienic gymnast, whom Brandt, however, far sur-
passes in knowledge, but from whose midst he has emerged, in
whose views he has grown old, and according to whose principles he is inclined to explain everything.

These four watchwords, however, have to be tested scientifically by means of all our physiological attainments. Empi-
ricism does not give any proofs. It attributes potent curative
factors to homeopathy also, although its basic principle (similia
similibus curantur) has been refuted long ago and its other rules
conflict with physical science. But this test need not be made
by a specialist; it belongs, together with hydrotherapy, balneology
and climatology, to a special teacher such as we have already in
Munich, and I hope other colleges will soon follow our example.
We cannot reasonably expect a gynecologist to give instruction
on the benefits obtained at watering resorts and climatic sana-
toriums nor would it be fair to expect him to teach the effects of

hygienic gymnastics, if, like climatology, gymnastics do not constitute an essential factor of his treatment.

However, I do not intend by any means to place hygienic
gymnastics and homeopathy on the same footing. I have
occupied myself but little with the former, but nevertheless
gained the impression that powerful curative factors are contained therein. Lay practitioners always overlook the fact that
a great many ailments are cured by the " vis sanatrix naturae "
without our assistance. To what extent the kinesiotherapist is
aided by autosuggestion, or how much he suggests to the
awakened patient, must also be taken into consideration. A
great deal of chaff will be blown away by an earnest scientific
investigation, but many a grain of gold will be left behind. I
myself have made a beginning in this matter, in so far that, in
knee-parting (Knietheilung), which is supposed to strengthen
the pelvic floor especially, and to cure cases of prolapse, I controlled the muscles of the pelvic floor by the sense of touch,
and, not noticing any contractions, I doubted the value of these
movements.

Dr. E. Ries, of Strassburg,* has given us very valuable
information by examining the possibility of diverting the blood
from the pelvic organs by means of gymnastic movements applied
to muscles remote from the pelvis, of conducting blood to the
pelvic organs by gymnastics used on muscles near or in the pelvis. The surprising result was that the gymnastic movements
always diminished the temperature in the vagina, and that consequently all the movements should be considered as diverting
the blood from the pelvis.

Brandt treats amenorrhœa and dysmenorrhœa only by
movements conducting the blood toward the pelvis.

*Dr. E. Ries on the Value and Significance of Gymnastics in Connection
with Massage According to Brandt's Method. Deutsche med. Wochenschrift
1892, No. 18.

Apart from these, it is only in incontinence of urine that these conducting movements are used as general treatment, in addition to compression of the sphincter vesicæ as local treatment. In all inflammations the general treatment is blood-diverting. Now, amenorrhœa having different causes, I treat the same causally, following good old medical principles. Besides, I always found pathological changes in or about the uterus, after removal of which dysmenorrhœa, with but few exceptions, disappeared with all its accessory symptoms. A few cases of incontinence of urine I cured by simple compression of the sphincter vesicae; in another case, where ischuria had remained after hemiplegia, compression had no effect in spite of movements conducting the blood towards the affected parts. For this reason I discard blood-diverting gymnastic movements in inflammatory affections of the pelvic organs, although, as previously stated, they hasten the disappearance of reflex symptoms. However, the above mentioned movements are discarded, not because they do not give good results, but because their benefits are out of proportion to the time and labor involved. I mentioned this in 1888, and I am of the same opinion to-day.

Only in cases in which it is especially desirable to aid local treatment, hygienic gymnastics must be considered the best form of treatment. Such cases I send to the Medico-Mechanical Institute in this city (Munich), to undergo a course of treatment by blood-diverting and respiratory movements. I have always preferred to treat my cases twice as long locally, because, by so doing, I was enabled to follow the local improvement more closely from day to day, and I was very much pleased to see how the reflex symptoms also gradually disappeared, while hygienic gymnastic movements grew exceedingly tiresome for me. The same has been the experience of a great many other physicians, and I remember well with what regret Brandt used

to tell me that his only pupil of earlier years, Dr. Nissen, of
Christiania, had given up hygienic gymnastics entirely or at least
to a great extent.   The so-called self-movements (Selbstbeweg-
ungen), that is movements which the patient can perform herself
without the aid of another person, are very useful, and I have
employed them very extensively up to the present time.

Even Brandt seems to admit their usefulness, for in his
recent publication* he gives an assortment of such movements
for practitioners.   If now a prominent gynecologist, in his inaugural dissertation,** avers that he has simplified, hence improved, Brandt's treatment by omitting hygienic gymnastics,
he is somewhat behind time with his views.   I have occupied
myself largely with gymnastics, especially with their scientific
aspect, in order that hygienic gymnastics might retain, or better
attain, that importance among physicians to which they are
entitled, namely that of an independent potent healing method.
For this reason I have described the single movements less
minutely than in my previous work, besides, Brandt himself has
devoted a good deal of space and illustrations to this part of his
method.   The physician must learn to recognize hygienic gymnastics as an independent remedial agent, and not as an appendage
to gynecology.   When, at the meeting of the congress of gynecologists in Freiburg (grand-duchy of Baden), I demonstrated
Brandt's treatment of prolapse of the uterus with all its gymnastic attachments, the unusual sight provoked general merriment
and shaking of heads among the attending gynecologists.   Nor is
it any less surprising that in Sweden, the very home of hygienic
gymnastics, the gymnast is struggling much more for recognition
than anywhere else, when we learn that the gymnasts since the

*Thure Brandt, Zur Massage der Prostatitis. Deutsche med. Wochen-
schrift, 1892, Nos. 44 and 51.
**Timling, Ueber Massage, insbesondere Dehnung und Loesung der
abnormen Fixxationen des Uterus und der Adnexe. Inaug. Dissertation, 1893.

time of Lingg find themselves in a similar relation to the medical practitioners as the physiopaths to the physicians in our
own country.   This may be to some extent the fault of the physicians themselves.   The Swedish physician studies for twelve
years, and it is his ideal to be a specialist in every branch of
medicine.   Herein lies the reason why he is so clannish profes-
sionally and not easily accessible to innovations.   Whether this
ideal can be attained at all, and whether it is practical to make
in that way medical service dearer in a country of comparatively scanty resources, need not be discussed in this work.
But even in that country a great deal has been done to put gymnastics on a scientific basis, physicians having been assigned to
positions in the Central Institute (the labors of Professor Mur-
rey, Professor Levin and Dr. Weil).   The director of the above
institute is not a physician, but an officer in the Swedish army.

Machine-gymnastics, invented by Dr. Zander and lately
revised by Dr. Levestin, of Stockholm, is in my opinion the
same in principle.   The machine has its disadvantages, lacking
the personal, cure-suggesting influence of the gymnast.   On the
other hand, the machine always makes the same movements,
does not become tired, and resistance can be measured more
accurately.   The personal influence of the superintending physician may also have an encouraging influence.   Of course, with
a machine we cannot produce all those manifold movements
which a human brain may invent or which a hand can perform; and manual gymnastics can never be entirely dispensed
with when gaps are to be filled which are inherent in machine
gymnastics.

LOCAL TREATMENT.

Local treatment consists, with but few exceptions, in massage and stretching. Massage serves to promote absorption of swellings and effusions and to relieve painful and swollen organs. Stretching is used in inflammatory processes where shrinking has set in, to lengthen the adhesions which have formed in the pelvic connective tissue, and to separate organs which have become adherent to each other by peritonitic exudations. It often happens that sensitiveness and swelling re-appear on account of stretching. In such a case it frequently becomes necessary to use alternately massage and stretching, mostly even during one sitting; but sometimes it is expedient to employ both at the same time. With the fingers of one hand in the vagina we stretch the fixating cord, while with the other hand the band is massaged from the abdominal wall. Having ascertained the nature of the patient's ailments and the latter pointing to a physiological disturbance of the pelvic organs, we take up the history of the case as a means to establish the cause of the disturbance; often it is not until then that we learn the whole extent of it, as certain physiological functions sometimes fluctuate within wide boundaries. Thus, one woman may bleed half a day during menstruation, another ten or twelve days; one menstruates every four weeks, the other every three weeks; in rare cases women do not menstruate at all. If no complaints or symptoms are manifest, if the type of the sexual life is not changed, if conceptions and births take place, all these various phenomena are within physiological limits.

Not until now should we proceed to a local diagnosis, the main part of which consists in bimanual palpation of the pelvic organs. All other methods, as inspection, probing, etc., must be used only when bimanual palpation yields insufficient information about the possibility of a pathological lesion lying at the bottom of the physiological disturbance.

DIAGNOSIS.

As soon as I became acquainted with Brandt's manner of diagnosing, in the fall of 1886, I tried it, and again and again compared his peculiar ways with the diagnostic method I had made use of up to that time. If I have adopted Brandt's peculiar methods to a great extent, the inference may be drawn that they must possess real advantages. I had a teacher in diagnosis whom every one will recognize as an authority. The frequent agreement of our diagnoses and the confirmation of the latter by operations, as well as the coincidence of my diagnosis with the pathologic-anatomical researches made later on, justified me to place full confidence in my diagnostic powers. I believe, however, I have earned the gratitude of many a patient because while examining her according to Brandt's method I spared her needless exposure, pain, and even narcosis.

The peculiarities of Brandt in diagnosis, as well as in massage, are as follows:

(1) The patient's dress is not removed, not even thrown back, but merely opened around the waist. The corset likewise is loosened, so that no hook or band may interfere. The chemise is then pulled up so far that the hand can be placed upon the bare abdomen; the abdomen itself, however, is not uncovered.

(2) The finger to be introduced into the vagina, from underneath the knee of the side corresponding to the hand employed, can also be advanced beneath the dress towards the vaginal orifice without the knees being separated.

(3) Only one finger is introduced under all circumstances,

preferably the forefinger, except in ventro-vaginal-rectal palpation, where the forefinger is inserted into the rectum and the thumb into the vagina.

(4) The hand laid upon the abdomen feels its way towards the finger in the vagina, not with uniform pressure, but penetrating deeper and deeper by means of gentle circular massage movements.

(5) The examiner, seated upon a chair at the end of a couch, takes the corner of the latter between his separated knees.

(6) Only a low bench, couch, or so-called plinth is used and no examining chair or table.

(7) The unemployed fingers are not flexed (examination with closed hand), but rest loosely extended in the groove between the nates (examination with open hand).

Of these special rules of Brandt, I have not adopted No. 3, 5, nor, except in rare cases, No. 6. All others I combined advantageously with my previous methods. No harm can be done to non-virgins by introducing two fingers, nor possibly any inconvenience caused. I, myself, with two fingers, reach about one half inch higher, and have the advantage of feeling stereometrically, being enabled to estimate distances by the distance of the two fingers from each other. In stretching, too, with two fingers I gain a better hold on the vaginal portion of the cervix than with one finger. In virgins, however, we must try to get along with one finger. In the latter I examine and massage by introducing the forefinger into the rectum, unless the diagnosis indicates that the hymen, for the sake of intrauterine treatment, cannot be preserved. In such a case I prefer the cleaner ventro-vaginal massage. The position of the examining physician (see Fig. 1) at the lower edge of the couch has only one advantage, namely, that of being the only

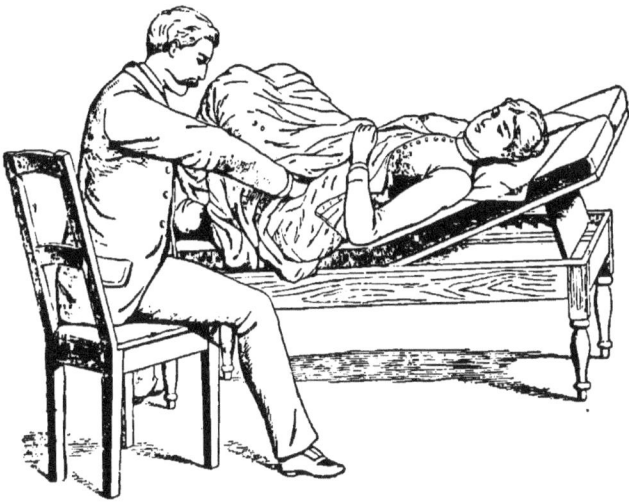

Fig. 1- Massage according to Brandt's method, the lower left corner of the couch between the spread-out knees, etc.

one in which the spine of the examiner is entirely straight during the examination or during massage; otherwise it presents only disadvantages, as I have found out by my own experience. In taking a position at the end of the couch, one has to reach too far up, impairing thereby the delicacy of touch. The position with the knees separated is likewise inconvenient, on account of the patient's dress; it is inelegant, and with ladies almost impracticable. It is not the position which cures the patient, but the correct appropriate movements of our fingers. For this reason I discarded No. 5 as unsuitable, seeing that the only advantage to be gained, "sparing of the spinal column," could be attained better by changing the position somewhat from time to time. Moreover, fineness of touch increases if one seats himself facing the patient, but further up, at about the latitude of the patient's pelvis. In regards to No. 6, the particular couch for the patient, Brandt told me that previously, while practicing in Sofge, a provincial town of Sweden, he had used only the plinth (bench). When I was with him, he used a short couch (couchette) constructed according to his own ideas, but only on account of its greater elegancy. I had my own couch (chaise lounge) built of such a height that it reached up to my knee and by a movable cushion could be adapted to the size of the patient. This arrangement I used up to 1888. At this time, however, I had a plinth constructed, i. e., a bench, the top of which could be partly raised so as to elevate the upper part of the patient as I had seen it done by Brandt. If the body lies upon the straight bench it is "wholly recumbent" (ganzliegend); if the top is raised and the upper part of the body elevated, we call it "semi-recumbent" (halbliegend). When the legs are flexed, the position is termed "relaxed semi-recumbent" (krummhalbliegend). At that time I still used the position with separated knees, at the lower edge of the bench. As my trowsers often became soiled by the patients' shoes, I purposely had the lower part of the plinth constructed in such a way that it could be raised. Subsequently I noticed that this change had insured two further advantages, the first of which enabled the physician to support his elbow during examination or massage, thus materially aiding his sense of touch. Through the second advantage the feet were comfortably supported by the elevated lower part and the abdominal wall of the patient relaxed, which favored diagnosis and massage. I do not pride myself on this improvement, but I thought it strange that Timling (l. c.) said Duehrsen had improved the plinth in the above manner. The plinth was constructed in this shape in the autumn of 1888, sketched in the spring of 1889, and described under Fig. 1 in Volkmann's Klinische Vortraege, old series, 353–354, January, 1890. Another improvement may be made, if the center piece, which, although the other two flaps are raised, remains horizontal, is made to slope toward the head piece, thus allowing the intestines to gravitate further upward, and thereby further facilitating palpation.

My experience justifies me in recommending this plinth * as a very practical piece of furniture for a physician. I have used it not only in massage, but also in plastic operations on the perineum, amputations of the cervix, Emmet operations, etc. Examination with the open hand seemingly not giving me any advantage, I adopted this method only recently, because I did not like the unusual abduction of the fingers. Only after repeated complaints of my patients that the knuckles of the closed fingers produced pain in the anal region, did I use massage with the open hand in order to spare the patients these pains; and since I have become accustomed to this modus operandi I believe I can reach thus even higher up than before.

* Measurement: Heighth 50 cm., breadth 50 cm., length 150 cm., head-piece 60 cm., middle piece 60 cm., foot piece 30 cm., supports at the flaps 20 and 10 cm.

COURSE OF EXAMINATION.

The course of examination is as follows: The patient places herself upon the plinth, both head and foot rest being elevated. She supports her feet upon the foot rest and is asked to move the perineal region as far forward and downward as possible. The pelvic incline is thereby diminished as much as desirable, the introitus has to come forward to its fullest extent, and the spine should not be curved, but should lie firmly and flatly upon the couch. The physician takes his place, as mentioned before, at the side of the patient, and with the fingers, which previously have been carefully washed but not dried and on which sufficient soap has been left to lubricate them, approaches the orifice of the vagina. It is best to flex the fingers (fold the fingers into the palm of the hand) while approaching the ostium vaginæ from under the patient's knee, in order to avoid soiling them. While formerly it was customary to press the fingertips against the perineum and to let them glide forward, until one became aware that they had passed over the frenulum labiorum, this method has been abandoned, lest germs, oxyuri, etc., might in this way be transmitted from anus to vagina.

To proceed from the front backward must be equally carefully avoided, as the touching of the clitoris or the urethra might produce pain or sexual irritation. At most, lateral movements should be made in order to find the vaginal orifice. In gliding over the frenulum into the vagina, we press the perineum strongly backwards; and as many women timidly retract the introitus vaginæ by enlarging the pelvic incline, we ask the patient to raise the buttock, thus securing for the fingers not introduced into the vagina a better resting-place against the sacrum.

In pressing the perineum back we are enabled to penetrate further and higher into the pelvis, as a glance at the sagittal section will show, and also avoid the irritable neighborhood of the clitoris and urethra. The thumb is now feeling for the perineum in search of any defects, especially of smaller ones, which otherwise might easily be overlooked, as the fingers did not glide over the perineum. This done, the thumb is placed sideways in one of the crural folds. During the introduction of the fingers into the vagina we should observe the condition of the hymen, also that of the columnæ rugarum (the transverse ridges extending outward from the raphe or columnæ of the vagina on either side) noticing whether these structures are free from scars, cysts, or tumors, and then proceed systematically to the spines of the ischium, keeping the palm of the hand still in sagittal position. Unmethodical examination makes a bad impression on the patient and gives unsatisfactory results. As far as the selection of the hand is concerned, it is very desirable to use both hands with equal ease and skill. However, if preference is to be given to any one hand, it should be to the left hand (except in confirmedly left-handed persons), as on account of its less frequent use it is softer, cleaner and more sensitive. Thus the effort made in counter-touch and massage falls naturally to the lot of the stronger right hand. After we have found the spinous processes of the ischium, which task is easily performed by gliding up and down the lesser sacro-sciatic ligament, we turn the exploring surface of the fingers towards the front.

If the vaginal portion of the cervix is in its right position, i. e., a little to the left of the center of a line drawn from one ischial spine to the other, it immediately comes in contact with the finger tips of the examiner; if the vaginal portion of the cervix is situated behind this line, it is usually due to contracted, double-sided posterior parametritis; if located anteriorly, or better in front, it is caused in most cases by retroflexion or prolapse. In a right or left displacement of the cervix we have to deal with one-sided parametritis in a state of contraction; but dealing with a dislocation towards the right, we have to take into consideration the fact that the viginal portion of the cervix lies naturally to the left of the median line.

We now move the fingers around the circumference of the cervix, trying to find out whether it is two-lipped or cone-shaped, i. e., if a birth or, more precisely speaking, a laceration has occurred, or whether there are any scars or cysts in the vaginal vault. With the index finger we then try to enter the external os, ascertaining whether it is wide or narrow, or looking for an ectropium, so-called erosions, cysts, or new growths. However, vaginal palpation is still incomplete, the very important task of ascertaining the mobility of the cervix remaining. One finger is placed to the right, the other to the left of the cervix, moving the latter alternately to the left and right side. By means of this manipulation we may detect any variation of the cervix, whether it is fixed to one side or another. *In case of unlimited mobility and healthy parametrium we are able to press the uterus to the left and to the right (on either side) against the wall of the small pelvis, this organ offering no resistance to our movements and without producing any pain. Mobility is anatomically as well as clinically greatest in entirely healthy genitals.*

In connection with this a law may be mentioned, which, as far as I know, was given by *Schultze:* If pain arises upon the side the tissues of which are stretched, we have a case of still

flourishing parametritis; if the pain is on the side towards which
the uterus is moved, we have to deal with ovaritis or salpingitis.   There are exceptions to this rule, but in most cases it
holds good, even when, as is usually the case, parametritis and
inflammation in the annexa are simultaneously present.   In this
instance pain arises by stretching as well as by compression of
the tissues on the affected side.

Not until now the right hand, which hitherto did not touch
the abdominal walls at all or at most only rested lightly in situ upon
the abdomen, so as not to interfere with vaginal examination,
gropes its way through the abdominal wall towards the fingers
of the left hand.

Not until now begins bimanual diagnosis.   The fingers of
both hands meet at first, without any tissues between them except
the abdominal wall, in order to become accustomed to the
touch-impression imparted by the latter.   The wrist of the left
hand is lowered as much as the couch will permit, while the tips
of the fingers feel upwards.   The right hand is placed upon the
inguinal region, or at most upon the iliac, and presses with
gentle rubbing movements (without, however, thereby displacing
the fingers upon the abdominal wall) this part of the abdominal
wall towards the median line, endeavoring at the same time to
touch the tips of the intravaginal fingers.   It is not advisable to
place the hand upon the middle of the abdomen and to try to
penetrate the fleshy recti muscles, pyriformis, etc., in a locality
where the panniculus adiposus is most abundant.   Thus even
in corpulent persons we will be able to make the diagnosis as
well as to apply massage in this way.   Not until having succeeded accordingly, in meeting the fingers of the left hand, may
we proceed with the examination.   Should we, however, be
unable to gain our point in this way we may relax the abdominal
wall by asking the patient to open her mouth and to breathe

quietly, or by diverting her attention from the examination by conversation; sometimes simply by leaving the right hand quietly in its place and asking the patient to allow it to penetrate deeper. Having attained our purpose, we place one finger into the anterior and one into the posterior vaginal vault against the vaginal portion of the cervix and try in the same manner to get the body of the uterus between our fingers. Then we examine the contour of the uterus, ascertain its size, its degree of softness or hardness, whether there are any tumors on it, whether the organ is sensitive, flexible at the point of flexion, or rigid (wink-elsteif), and especially whether the fundus can be felt anteriorly or posteriorly. However, we cannot rest satisfied with mere absence of the fundus in the anterior vaginal vault, as it may be found also in the pelvic axis; nor is the presence of a tumor in the posterior vaginal vault a sufficient indication, as this may be a new growth. Only absence anteriorly and presence poster-iorly furnish us an unerring diagnosis of a retrodeviation. In rare and abnormal cases sounding may be required for determin-ing whether it is a fibromyoma or the uterus.

If retroflexion is present, it is part of the diagnosis to replace the uterus if possible and to test its mobility. In such a case we try to move it not only from side to side, but also back-wards and upwards. *Retroflexions are most frequently caused by contraction of the pelvic connective tissue in front and extend-ing in the direction of the obturator foramen*, occasionally also by contraction in the direction of the spina ischii sideways and towards the front. In cases still more rare this disease is pro-duced by parametritis superior, i. e., by shrinkage of previously inflamed parts, extending along the spermatic (ovarian) vessels from the cornua of the uterus towards the kidneys, thereby dragging the uterus first sideways and then upwards towards the back. *Peritoneal adhesions are not the cause, but the conse-*

quence of retroflexion. To distinguish different forms of retroflexion, as is still done, or to draw a line between retroflexion
(as a grave disorder) and retroversion (as a mild one), is idle
play. For it has been demonstrated that retroversion presup-
poses a rigid metritic uterus, and that in retroversion of the first
degree, the greatest complaints often exist, whereas, on the
other hand, a flexion with a non-rigid uterus is a sign of minor
affection of the uterine wall and retroflexion of the third degree
may run a symptomless course. The term "retrodeviation"
would therefore be more suitable. The uterus having been
examined, we proceed to the examination of the tissues surrounding this organ (parametrium). (See Fig. 2.) I use two
methods in order to find the ovaries. Either I depress the wrist
of the hand in the vagina as forcibly as I can for the purpose of
reaching with the finger tips as high and as far to the back as
possible, allowing the tissues to glide inch by inch through my
hands while groping with the external hand towards the inner
one; or I begin at the cornua, allowing the external and internal
fingers to meet again and examine sideways and towards the
back. The distance of the ovaries from the uterus differs very
much in individual cases. Thus, in one person I found the left
ligament of the ovary more than one centimeter longer than the
right ligament. The left ovary is most easily examined by using
the left hand, and the right ovary by using the right hand for
internal examination. In corpulent persons the external hand
is placed upon the *right* groin, and the abdominal wall, which
is thinner in this region, is drawn over the ovary. The right
ovary lies, corresponding to the physiological slight dextrotorsion
(torsion towards the right) of the uterus, in most cases a little
further posteriorly. Having found the ovary, we test its mobility by drawing it forward in the direction of the corresponding
(right or left) oblique diameter of the pelvis. Normally movable

ovaries can be brought near to the horizontal ramus of the pubis of the opposite side. Should the ovary be fixed and painless, it is due to *old* periovaritis; if it is fixed and painful but *not* enlarged, we have to deal with *acute* periovaritis.

The normal ovarian tubes can be found only under specially favorable circumstances; on the other hand, a tube which is hardened and enlarged by chronic inflammation is easily recognized by its winding course. Very often, however, we have to content ourselves with simply having felt a tumor beside the uterus, and only during the course of the massage treatment can we decide definitely whether it is a tubal tumor, an ovarian tumor, or an apparent tumor ("conglutination tumor"), the latter produced by peritonitic agglutinations of the ovaries with the overlying tubes, and perhaps, also, with overlying intestinal coils. Tumors of the last category often break up during massage by disintegration into their constituent parts. As long as the agglutinated intestine does not yield, it may at times be felt quite distinctly.

In conclusion, we have to examine bimanually the parametric band that betrayed itself by the fixation of the cervix, by the extramedian position of the vaginal portion of the cervix, or by the characteristic symptoms, the presence of which we suspected while making the mobility test. This band is detected most easily by stretching it as in the "mobility test," but only with one finger, because the other has to be used for countertouch. It then becomes an easy matter to find out its thickness, firmness, etc. The physician who does not trust his sense of touch, but would like to see the cord itself, can easily do so by introducing a Sims speculum in the knee-chest position, seizing at the same time the cervix with a vulsellum, and moving it towards the other side. The cord may then be seen projecting beneath the vaginal wall, laterally and posteriorly in case we have to deal with posterior parametritis (so-called anteflexion); but it projects upon the other side, or better, the lateral vaginal wall, if we have before us a case of retrodeviation.

I have dwelt so extensively upon the diagnosis, because the diagnostic procedure is influenced somewhat by Brandt's method. Moreover, a much more accurate diagnosis is necessary in regard to certain matters, as for instance the fixations, in order that the method may be profitably applied—quite an important factor, which Schultze already pointed out in his preface to Profanter's book. *This, however, ought not to deter the less experienced, because there are no means by which one can acquire diagnostic dexterity so quickly and so surely as by the practice of Brandt's method.* The beginner has only to be more cautious than the experienced.

In order to present the above diagnostic procedure in a condensed form, stripped of all needless verbiage and lengthy explanations, I give the following resumé:

(1) Approach the ostium vaginæ with the fingers from the side, and examine the perineum with the thumb.

(2) Pay attention to the condition of the introitus vaginæ, the hymen, and the vaginal wall, while inserting the fingers and pressing down the perineum.

(3) Search for the spines of the ischium, turn the finger-tips forward, and ascertain the relative position of the cervix to the median line, as well as to the spinal line.

(4) Palpate the cervix and examine its form.

(5) Try to enter the external os, ascertain its size and the character of the mucous membrane.

(6) Try to press the cervix to the right and to the left against the pelvic wall, also against the sacrum.

(7) Bring the finger-tips of the two hands together, one

hand examining from the vagina, the other from the abdominal wall. The hand is placed at first upon the inguinal region.

(8) Ascertain the position of the body of the uterus— whether it lies anteriorly, posteriorly or in line with the pelvic axis, by placing one finger in front of the cervix, the other behind; making at the same time counter-touch from the abdominal wall. Thereupon examine the form and size of the uterus.

(9) Search for the ovaries, ascertain their size, form and mobility; locate, if possible, also the tubes and examine any parametric cords by stretching them bimanually in regard to their firmness, length and thickness.

MASSAGE.

Massage of the pelvic organs is employed almost exclusively in the form of circular rubbings, only in special cases, as in the process called " malning" (painting), and to be described hereafter, in the form of stroking movements. In principle there is really only a slight difference between gynecological massage and general massage as used on other parts of the body, because in making circular rubbings we do not press equally upon all parts, but increase the pressure at one place of the periphery of the circle and possibly in the direction of the venous and lymphatic circulation. This object may be obtained very easily if, in massaging with the right hand upon the left side, we describe a circle from left to right, and, in massaging the right side, describe a circle from right to left. The intravaginal fingers do *not move*, but simply lift the organ to be massaged so high that only a little soreness is caused, but no real pain. The external hand, which has to perform the massage, is placed very lightly upon the abdominal wall and describes large circles in the beginning; gradually the intestinal coils give way and the hand penetrates deeper and deeper into the pelvis. The circles become smaller the deeper the hand advances, and depend finally upon the size of the organ which we intend to massage. When, for instance, we massage an ovary, we often describe circles of smaller circumference than the ovary itself. Such a manipulation is apt to convey the impression that the hand of the operator is trembling. A sitting lasts ten or fifteen minutes, and may tire the beginner very much. Soon, however, he becomes accustomed

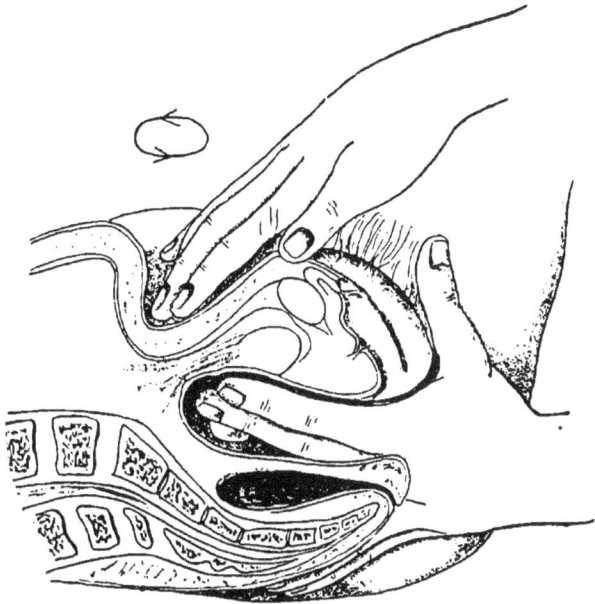

Fig. 2. " Massage of perimetritis chronica dextra."

to it, being able to massage ten or fifteen persons in one fore-
noon. In concluding the massage we do not stop abruptly, but
relax the pressure gradually by making the circles larger and
larger. The force of the pressure used during the sitting depends
in general upon the sensitiveness of the patient, just as the intravaginal hand, so the external hand should produce only slight
pain (*a sore feeling*). In using too much pressure we cause
unnecessary pain, sometimes even an aggravation of the present
inflammation. If we exercise too little pressure the patient does
not derive sufficient benefit from it; moreover, a condition might
be produced, for which, by the way, massage has been unjustly
reproached, i. e., sexual irritation. The latter, however, may
easily be allayed by increasing the pressure a little, so that real
pain is caused. The other manipulation of Brandt, compression
of the pudic nerve, I never found necessary for accomplishing
this purpose. The posture of the patient, the position of the
physician's fingers and the arrangement of the hands upon the
patient's body are the same as indicated above in describing
the diagnosis. As in the latter, so also in massage, the external
hand does not displace the skin, but the hand and the abdominal
wall together rub the diseased organ. If we follow the rules
given above sexual irritation will occur no oftener during massage
than during any gynecological diagnosis. No doubt there are
women of abnormal sexual excitability, who become excited
by every gynecological examination, sometimes even without any
such cause. Those malevolent persons, however, who times and
again raise such an unwarranted objection against the method,
certainly without fully understanding the same, on the one hand
unjustly charge the numerous physicians using this method either
with carelessness or unscrupulousness, it being an easy matter
to find out whether a woman is sexually excited or not; while, on
the other hand, they most grossly insult thousands of highly

respectable women by maliciously insinuating that they undergo
this treatment in spite of its power to cause sexual excitement,
or, perhaps, even on that very account.   That Englishman, who,
in the first edition of Nebel's book, "Ueber Bewegungskuren"
(On Movement Cures), is quoted as having said: "Massage in
gynecology is masturbation of women by a man," did certainly
not exert himself very much to acquire a thorough knowledge
of this method as a curative agent.

The rubbing movements of the external hand are executed
with the finger-tips (see Fig. 2).   Only in cases in which larger
surfaces have to be massaged, as for instance in a myoma the
upper surface of which is very sensitive, or in a case of large
retrouterine hæmatocele Brandt uses the whole hand, rubbing
with the palmar surface (see Fig. 4).   Frequently after two or
three sittings, a sense of formication is felt in the finger-tips,
and sometimes the delicacy and acuteness of touch may even
be temporarily impaired.   In order to avoid such occurrences, I
usually massage with the index finger of the slightly flexed hand
(see Fig. 3).   In this way I am also enabled to get my hand
behind the uterus as far down as the Douglas space without any
difficulty, and to massage between the uterus and the ovary; in
short, I can penetrate quite easily into the pelvis and massage
ovaries and uterus separately.   Of course, it would be impossible
to loosen a fixed ovary or retroflexed uterus by this manipulation,
the surface of the flexed finger not being pointed enough (see
Fig. 3).   This kind of massage, however, has the advantage
that it may be kept up for any length of time, and that it is
more agreeable to the patient, the abdominal wall not being
abused as much as by massage with the finger-tips.   It is advisable to use alternately the margin of the flexed fingers or the
finger-tips, according to personal preference or the advantage to
be gained, just as we may change our position in diagnosis, as

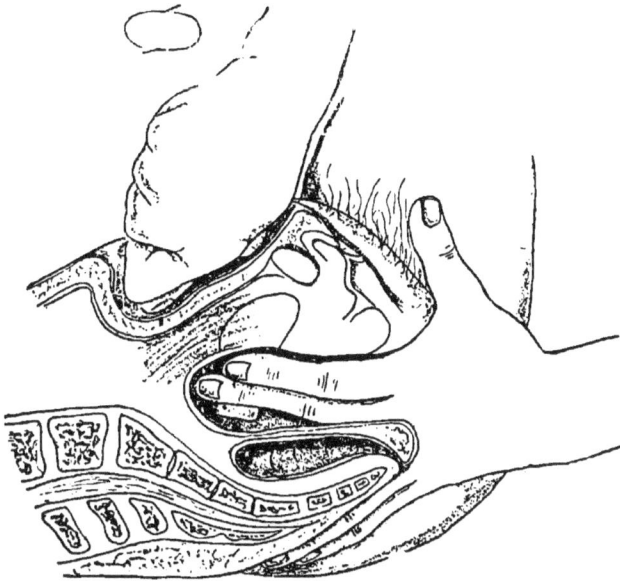

Fig. 3. Massage of parametritis chronica dextra (under slight pressure) with the index finger and left closed hand by the author, ⅓ natural size.

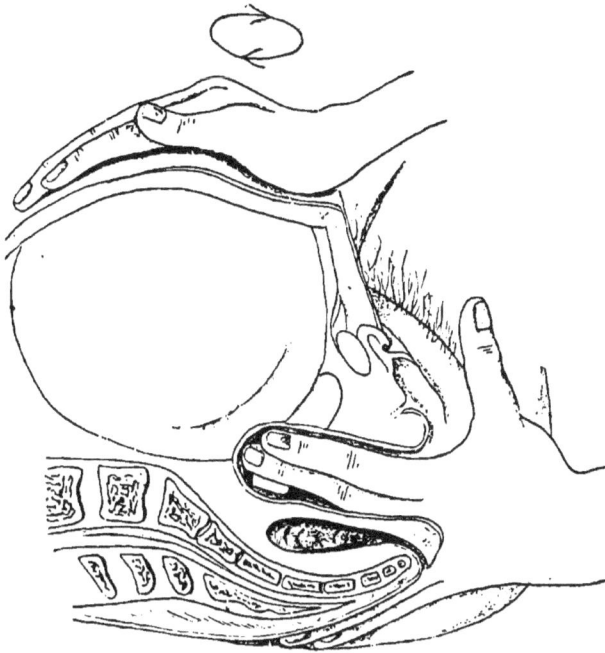

Fig. 4. Massage of larger surfaces with the ball of the hand (myoma, gravid uterus, etc.) according to Brandt.

well as in massage. Position, as I have said, is the least important factor, the patient being cured not by the latter, but
through the correct movements of our hands.

Contraindications for Massage.

Massage and stretching are contraindicated by "pus and
cancer."

These laconic words of Brandt are to be interpreted in such
a manner as to include under "cancer" every malignant new
growth, and under "pus" the presence of infectious micro-
organisms in the internal genitals.

Tuberculosis, actinomycosis and fresh gonorrhœa are con-
traindications for massage, as well as fever, a symptom indicative of acute septic processes in or about the uterus. On the
other hand, "pus"—that is, the liquefied product of suppura-
tive inflammation—does not necessarily contraindicate massage.
More than once I have seen pus of pyosalpinx discharge itself
without causing any harm, for the evident reason that the germs
ceased to be infectious—a condition which, according to the
investigations of *Prochownick*, may be brought about three
months after the primary infection. But in these very cases
the *greatest* caution is required. In every instance let us follow
the rule that, if during the first sitting the pain has not been
diminished, we have to be twice as careful at the next sitting,
because an insufficient massage can be remedied the next time;
whereas it is not so easy to correct any harm caused by using
too much of it. Another precautionary measure to be adopted
is, not to follow the course of the vessels, as is usually done in
tumors about the uterus the nature of which we cannot diagnose

at once, nor to massage away from the uterus, but *towards* it,
i.e., always in such a manner as tubal tumors ought to be massaged.   Pregnancy is no contraindication for massage.

Indications for Massage.

Massage is indicated in every case in which pressure on any
part of the internal genitals causes sudden and perhaps persistent
pain, or in which we want to disperse a swelling or effusion.

These pains are usually increased after the first sitting. This
fact must be borne in mind and the patient instructed accordingly, otherwise both parties might be kept from the continuance of
the treatment or the patients might be worried unnecessarily. A
cold application according to Priessnitz or an ice bag (Nissen)
almost invariably give relief; if not, a rectal suppository of
opium (grains $\frac{1}{2}$) will surely subdue the pain.   It is a good
prognostic sign if the pain diminishes during the first sitting,
even if it should return with greater severity.   Should the pains
decrease during the first sitting, they will be less severe after the
second sitting than they were before the treatment.   This improvement seldom fails to manifest itself after the fourth or
fifth sitting.   The most obstinate case to yield to our treatment
is chronic peritonitis.   Here (true only in rare cases) we can
press only so lightly for months that we suppose details rather
than feel them.

I never found that (by massage) pain was increased for any
length of time or that any permanent harm had been done.
Where fever is present, it is necessary that feverless intervals of
months' duration elapse before we dare begin the massage.

As special indications for massage may be mentioned:

Ovaritis, periovaritis, salpingitis, pyosalpinx, hydrosalpinx, chronic adhesive peritonitis, remains of exudations of retrouterine hæmatocele and fibrinous peritonitis. Combined with stretching massage is indicated in chronic parametritis; in short, in all chronic inflammatory processes resulting in shrinking and adhesions combined with subsequent treatment of the endometrium, it is also to be recommended in chronic metritis, subinvolution of the uterus, inflammatory processes on the pregnant uterus, paraproctitis, paratyphlitis, and perityphlitis.

OVARITIS.

The diagnosis of ovaritis, and especially the differential diagnosis between ovaritis and periovaritis, is not always easy. The ovary is sensitive in case of periovaritis, but not enlarged. If the ovary is not enlarged, but glued firmly to its substratum (base), we have to deal with a chronic adhesive periovaritis. In case the ovary is movable, however, we have to examine first the ovary on the other side, and only when the latter is of the same size but painless, can we assume the existence of periovaritis of the former. On the other hand, if the other ovary is smaller, we have to deal not only with an inflammation of the lining coat, but also with a parenchymatous inflammation. Of the four cardinal symptoms—redness, heat, swelling, and pain— we cannot see the first at all. Increase of temperature and pulsation are signs of minor value and often wanting. In regard to enlargement, we have to remember that the size of the ovary differs very much individually. I have in my possession a specimen $2\frac{1}{2}$ cm. long and another $8\frac{1}{2}$ cm., and both were healthy ovaries. Therefore the old rule, to compare both sides before we allow ourselves to speak of enlargement, holds good here as well as in surgery. One phenomenon, however, deserves to be pointed out especially, and that is the form of inflamed ovaries. This is nearly spherical. It is, furthermore, hard to determine whether pain is present or not, as the healthy ovary, too, is sensitive on pressure up to a certain degree. It stands about as much pressure, without real pain being caused, as the testicle of a man, both being supplied by the same nerves. But the pressure is hard to measure, the resistance of the abdominal wall and the endurance of the patient differing in a great many cases.

The complaints in ovaritis consist in pain in the region of the diseased ovary, radiating from the groin in the direction of the lumbar vertebræ. Not infrequently there is pain across the sacrum, and, on pressure, in the compressed ovary, but at the same time also on the opposite side. Likewise in ovaritis and periovaritis the pain almost always radiates toward the knee in the direction of the obturator nerve at the anterior part of the thigh. This affection might be called obturatorias, analogous to the term ischias (sciatica). Such pains are often the cause of confinement in bed for years. In other instances the patients are still going about, overcoming the pains, which are worse at one time and better at another. In these cases a great deal depends on the fact whether the patient is well or badly nursed, and whether she remains in bed permanently or is still able to go about. It seems paradoxical, but it is nevertheless true, that in no other cases will too assiduous attention do more harm than in the ones just described. When solicitous relatives are bent on continually reminding the sufferer of rest and endurance, and the over-conscientious practitioner, true to habit and tradition, is inculcating constant rest in bed, perhaps even rest in a recumbent position, the ovaries sink backwards, fibrinous plastic exudations are transformed into pseudomembranes and produce adhesions, which have sufficient time to become firmly adherent. On the other hand, the appetite is lessened by morphine and other hypnotics; sleep does not come to the inactive body, the consequence being loss of strength and energy.

It is different, however, when an energetic husband, or a business where the wife is indispensable, urge the latter to move about to a certain extent, or when her energy and desire to
4

enjoy life are such that she is either unwilling or unable to act
in strict accordance with medical instructions. *In this case*
there will be less waste of bodily strength and the disturbances
less severe, because the ovaries, if they do become fixed, are not
so closely adherent and do not sink back so far.

We cannot speak very well of ovaritis and periovaritis separately, because these two pathological conditions are for the
most part associated with each other, just as pneumonia rarely
ever occurs without accompanying pleuritis. The agglutination
of *two* peritoneal surfaces with each other is the consequence,
as I have found out by experience, of a previous inflammation of
each one of them. I am convinced of this fact, because we
may find enlarged, inflamed ovaries which are entirely movable.
How this inflammation, often confined to the ovaries, originates,
is still an unsolved enigma to me.

In some cases there was a history of masturbation, which
might have been the cause of this rather rare but usually bilat-
eral disease. But one must not be too hasty in generalizing.
In those cases, in which the affection of the ovary is due to an
inflammation of the uterus, we seldom fail to find also diseased
perivascular connective tissue of the vessels supplying the
uterus—so-called parametritis. This disease, too, rarely runs
its course without participation of the peritoneum, and so we
have two inflamed peritoneal surfaces, which are apt to become
agglutinated.

The TREATMENT consists in circular rubbings not extending
beyond, but limited to the inflamed organs; the pain itself is sufficient indication. The swelling of the ovaries subsides often so
quickly that one feels inclined to think of an erroneous diagnosis. On the following day, however, our diagnosis is readily
confirmed by the return of the swelling, which usually occurs

after the first sitting. This phenomenon can be explained only by the fact that we had to deal with œdema of the ovary.

In other cases, especially in isolated, double-sided inflammations, sensitiveness only disappears, but the enlargement of the organ remains almost entirely. This phenomenon must be accounted for by the formation of cicatricial tissue in the ovary, and the solid character of this tissue by the fact that masturbation, which may have been the possible etiological factor of such an inflammation, had begun at the time when the ovaries were still in process of development.

CURE OF SMALL OVARIAN TUMORS.

Should there be any cysts on the ovary, they can often be crushed and made to disappear permanently by massage. As I have been treating such cases for years, I can safely say "permanently." The rubbing of the epithelial surfaces upon each other in the vacant space evidently destroys their "hydropic" property. The largest cyst which I removed in this manner had a diameter of seven cm.; in most cases, however, they are smaller. In one instance the removal produced a sensation similar to that experienced in crushing grapes in the bunch. Larger cysts usually become soft because they empty themselves gradually; others burst suddenly. The largest cyst which disappeared upon massage had a diameter of ten cm. It was removed by Dr. Heinss, a pupil of mine, when he was still a beginner in massage treatment. Both having agreed at a previous consultation as to the presence of a cyst, Heinss massaged the latter, as well as the other inflamed pelvic organs. The cyst was massaged on account of its great sensitiveness. One day, however, it burst unexpectedly during the massage. While there had been no complaints whatever in any of the other cases, there was considerable pain in this particular one, also a feeling of oppression and for some days subfebrile temperature in the evenings. It would not be advisable, therefore, to attempt the removal of tumors larger than the one just mentioned. A tumor of this size might justly be regarded as the proper limit. In other cases I purposely tried to burst such cysts, but without success.*

At the time when Brandt was in Jena he claimed to be able to cure ovarian tumors. Schultze and his assistants, however, seriously doubted Brandt's ability to effect such cures, and when Schultze had convincingly and conclusively demonstrated the presence of a tumor which Brandt claimed to have removed, the latter henceforth, in his writings, carefully avoided all mention of a cure of ovarian tumors. But the fact that ovarian tumors may burst, had been established long ago by various operators, together with the proportionately small amount of danger resulting from absorption of the discharged contents.** I am enabled to contribute the following cases to the solution of this question:

Mrs. L., of S., consulted me in the summer of 1889. On examination I found an ovarian tumor of about seven cm. in diameter on the right side (see above), together with a retroflexion, laceration of the cervix, and left parametritis. I succeeded in removing the retroflexion, together with the parametritis, in a short time. The patient urged an operation for the relief of her tumor, because her daughter, who, for the time being, had charge of her business, intended to get married and to go to America, and the patient could not be operated on at a later date without great pecuniary loss. Two days before the operation I could not detect any tumor, and I told the patient the tumor had burst, but would most probably fill again. The

---

*In the meantime, my assistant, Dr. Schmidt, and I have removed a cyst measuring 12 cm. in diameter without any disturbance of the patient's health. Massage was employed only for the purpose of soothing the pain for a time and to persuade the patient to have an operation performed. The exact limit of a cure of ovarian tumors by massage so far as size is concerned is thus still a matter of uncertainty.

** Comp. Kuestner, Ueber Bilinurie bei geplatzen Ovarientumoren.

patient, however, wished to be operated upon. I removed a
cystically degenerated ovary, which, being deeply imbedded in
the broad ligament, was hard to extirpate, as it did not furnish
a good stump. The specimen, which is still in my possession,
shows a cavity in which a hen's egg could easily be placed. I
know of ten cases which have been cured. Two of them were
treated by my assistant, Dr. Schmidt—I simply establishing the
presence and, subsequently, the absence of the tumor; one by
Dr. Flatau, his predecessor; one by Dr. Heinss, in private practice, and six came from my private clinic. In one case a relapse
occurred after a year, a small tumor appearing on the anterior
and a large one on the posterior aspect of the original tumor,
and at the side a tumor measuring about $2\frac{1}{2}$ cm. Since that
time nothing else has appeared. I maintain, therefore, that the
cure of small and thin-walled ovarian cysts by massage is possible and without danger. At the same time I must say, however, that a tumor of 10 cm. diameter is about the limit for
massage. If this limit is exceeded, the method is as dangerous
as laparotomy. The latter is to be preferred in cases in which
this limit has been overstepped, because we can then see and
control the consequences of our work. An infection of the
peritoneum (Fritsch) by colloid masses, colloid degeneration of
the same, and dissemination of daughter-cysts upon the omentum,
etc., never happened in any case. The question is whether
there might not have been a colloid cancer in those cases. I
examined a case of colloid infection in which I could detect carcinomatous structure only in the most recent metastasis in the
liver. The epithelium in this case had already undergone colloid degeneration. Otherwise I found only colloid material and
a structureless yellow coating of the abdominal organs as the
remains of the carcinomatous tissue.

PSEUDOTUMORS.

—

If I said before that the treatment often had to aid us in making a diagnosis, that periovaritis combined with perisalpinitis, inflammation and agglutination of the serous coat of the intestines simulates a tumor, which, however, dissolves itself after a few massage sittings into its slightly enlarged constituents, I protest against any assumption that the above mentioned tumors might have been such as my assistants and I between ourselves are pleased to call " conglutination-tumors "—a term certainly not very fine, but nevertheless suitable. Tumors of this kind have mostly undefined limits, and if it were not for the great sensitiveness we would hardly fail to establish immediately the absence of a real tumor. If, in order to clear up our diagnosis, we resorted to anæsthetics, as has been usually done in such cases, we would find at once that there is no tumor; but in doing so we might fail to notice that the organs lying upon each other are agglutinated into an apparent tumor. Therefore, what I have said does not apply to such tumors, but only to cysts which persisted after more or less protracted treatment and have sharply defined outlines.

In case the ovary is enlarged, sensitive, adherent to its base, and its limits are vague and indistinct, we have to deal with *ovaritis* and *periovaritis*. Should we, on the other hand, find sensitiveness, fixation, and undefined outlines without enlargement, we have a case of pure *periovaritis*.

In both instances restoration of mobility is the most important part of the treatment. The anterior, straight border of the healthy ovary is attached to the broad ligament, the convex surface and posterior border are free and movable, and the organ stands, as it were, erect when we stretch the broad ligament. It is this free convex border which, in periovaritis, is held fast by pseudomembranes. To massage this border, to try to loosen it from its base, and to pull it forward, is the most important task to be performed. In order to accomplish this, we often have to depress the wrist of the left hand as deeply as possible and to ask the patient to take the most favorable position imaginable (see above), so that we can reach as far backwards and upwards as possible (Fig 5). The right hand, with the finger tips directly towards the pelvis and the wrist towards the region of the spleen, is placed upon the abdomen in the region of the left groin (supposing we have to deal with the left ovary, which is usually the case). We then try, by massage movements, to reach behind the pathologically fixed free border of the ovary, and while the fixating pseudomembranes are gradually stretched and rubbed away we pull this border in the direction of the oblique pelvic diameter forwards, thus endeavoring to loosen it from its base.

Schultze has loosened fixed ovaries for years, but with the help of anæsthesia. I myself often administered the anæsthetic while *Schultze* employed this method of loosening the ovaries, and on one occasion he told me that since the operation the patient, who had been confined to bed for years, had experienced no more trouble in walking, and that he could proudly look back upon a series of similar experiences. The principle, accordingly, is not a new one, and I myself at various times resorted to anæsthetics in loosening the ovaries according to *Schultze*, and subsequently removed the inflammation, soothed the pain, and prevented relapses according to Brandt.

In all cases in which the pain radiates along the course of the

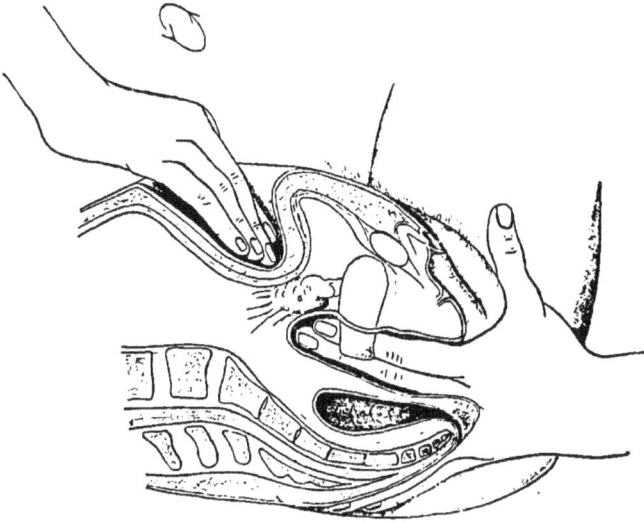

Fig. 5. Massage and Loosening of the Ovarium. Sketched by the author.

obturator nerve (see above), I prescribed hygienic gymnastic movements, to be performed at home by the patients themselves, and which may be designated as foot-instep-standing (fussriststehend), that is, genuflexion with arms akimbo. The patient stands in the room, arms akimbo. Behind her is a footstool at such a distance that she can barely reach it with the tip of the painful leg, which is stretched far backwards and rests upon said stool. The patient then lowers herself with arms still akimbo and genuflexes as far as she can, thereupon raising herself again. It is essential to tell the patient in unmistakable language how often and how the movements are to be made, as otherwise she might regard them as immaterial and not follow the instructions. About ten times morning and evening will be sufficient. Distance and height of the foot-stool may be increased after a time. In the beginning it is often necessary to stand in front of the patient on a chair and to assist the enfeebled woman, who is sometimes unable to raise herself again unassisted, whereas later on resistance may be used to increase the effort made by the patient in rising. By this movement the obturator nerve is stretched and becomes loosened from adhesions and constricting scars. It is analogous to another movement which I saw executed in the Central Gymnastic Institute in Stockholm for the cure of sciatica. Although this disease appears only as a symptom in posterior parametritis I nevertheless shall describe on this occasion the gymnastic movements used for the cure of sciatica.

The patient afflicted with sciatica stands in front of an inclined plane. In Stockholm it was a long beam with indentations, about half a meter (about 1 foot $7\frac{1}{2}$ inches) apart from each other, which served as a support for the heel of the exercising patient. The latter lifts her foot until pain ensues, and supports the heel in the notch suitable for the severity of the pain and the degree of her sensitiveness. The patient now bends the upper part of her body forward as much as possible with arms akimbo, and rotates it about ten times to the right and as many times to the left. She then lifts the foot up to the next notch and performs the same movement until pain warns her to be cautious. In this way the sciatic nerve is stretched across the neck of the femur on the same principle as the nerve-stretching after Nussbaum and as the previously mentioned movement for obturatorias. Many a case of sciatica have I cured by this movement. It is also a valuable agent in the treatment of pain in the obturator nerve.

SALPINGITIS, HYDROSALPINX, PYOSALPINX.

It is best to discuss this group of diseases together, just as we grouped ovaritis and periovaritis. The nature of salpingitis determines the development of hydrosalpinx or pyosalpinx when it comes to occlusion of both ostia. Where we can make out a sensitive cord, extending from one of the cornua of the uterus toward the ovary, accompanied by colic-like pains in the side and a sense of oppression with or without fever, we are justified in assuming the presence of salpingitis. Only in chronic salpingitis do we find the tubes more resistent and the outlines of its windings more strongly defined than is naturally the case. There can hardly be any doubt that in most instances pyosalpinx is caused by purulent processes which have advanced from the cavum uteri into the tubes. I have seen cases in the autopsy room in which every cross-section showed the lumen to be filled with pus. Gonorrhœa takes the first place among these purulent processes.

The best plan to follow in cases of fresh gonorrhœal infection in which the patient complains of colicky pains in the side, is to abstain from a local diagnosis, as an examination might do immense harm by forcing the pus into the abdominal cavity. It is good policy to wait—even with intrauterine treatment—until the ostium abdominale becomes closed. Although gonorrhœal endometritis is a common disease, fortunately it does not often occur that the disease extends upwards, but is most frequently limited to the cervix, and, indeed, remains very often and for a long time confined to the uterus. As soon as we can determine

the presence of a painful tumor beside the uterus, which might
be either salpingitis or pyosalpinx, we must, even in making
our diagnosis, examine and press only *towards* the uterus.   In
using massage on the diseased tubes we ought never to forget
that every pressure should be made *towards* the uterus.   Now,
if the ostium uterinum has become occluded and pus has accu-
mulated in the tubes, it is often easy to make a correct diagnosis.   The peculiar form of the tumor, characterized by a
gradual decrease in size towards the ostium uterinum and
increase towards the fimbriæ and its tortuous and irregular dilatation, constitutes one of the main factors in diagnosticating
this disease.   The diagnosis becomes still more certain when
we find a *non*-enlarged ovary.   Should the ovary, however, be
found to be changed into a cystic tumor, the diagnosis becomes
more complicated.   Two cases of this kind I have had to deal
with up to the present time.   In one case the diagnosis pointed
either to an ovarian cyst or pyosalpinx, probably the latter.
Both conditions were found to be present.   In the other case
an ovarian cyst was assumed, but a flaccid pyosalpinx was
found with it.

The TREATMENT of pyosalpinx, as well as that of chronic
purulent salpingitis, consists in massage and gentle stroking
towards the uterus, special care being given to the straight part of
the tube.   This treatment, however, should not begin until at least
three months have elapsed since the infection, as the germs keep
up their virulence at least for that length of time.   (Prochownick.)   The fimbriæ have to be spared as much as possible.   It
is of prognostic importance to ascertain how far the lumen of
the pars recta tubæ is occluded, and what proportions the
pyosalpinx has assumed.   In case the occluded segment is long,
there is but a slight chance of making it passable again.   On
the other hand, if the tubal sac is large, the tube will in all

probability, in spite of its having been emptied, not contract
again to its former size, but a hollow space will remain, most
likely to become filled again. The largest tubal sac which I
emptied successfully measured about eight cm. in diameter.
However, it was not evacuated by massage, but only by introduction of a tent. The case was that of a girl, twenty-three
years of age, with a pyosalpinx of five cm. in diameter on her
right side, and one of eight cm. on her left side, both cysts nearly
touching each other behind the uterus. Extensive peritonitic
troubles disappeared under massage, and finally (after two
weeks) the tumor on the right side emptied itself after a
massage-sitting, while she was riding home on the street
car. Although I continued the massage for a week and
pressed quite energetically upon the only slightly sensitive
sac, the other, larger pyosalpinx did not empty itself. Thereupon I introduced a tent under aseptic measures and at the same
time made preparations for laparotomy. The tent remained in
place for twelve hours, and although there was no fever present
the pains were very severe. The tampons were soaked well
with pus, and about a tablespoonful of it escaped after the
removal of the tent. For some time afterwards the patient could
feel how during the daily antiseptic uterine douche the fluid
spread to right and left. This feeling and the purulent discharge disappeared gradually. The massage was afterwards
continued for a short time (about eight days) leaving the tubes
in a slightly swollen but painless condition. No troubles have
recurred since.

Up to the present time I have emptied about fourteen tubal
cysts by massage. To this number may be added nine cases
which were treated under my supervision by my assistant, *Dr.
Schmidt*, and his successors. In four of these cases the contents
were doubtless purulent. In the other cases an exact observa-

tion could not be made, tumors rarely ever emptying themselves
during the sitting. At the present time I am treating a lady
afflicted with a pyosalpinx of a diameter of about fourteen cm.
on the right side and one of four cm. on the left side. By
means of massage, continued for four months, I succeeded in
removing pus at the beginning, later on pus mixed with debris,
and since that time about nine times a more serous fluid from
the left tube. A friend of mine had charge of the case before
me. He treated only the larger tumor by massage, producing
quite often tubal colics and slight increase of temperature.
This right tumor I removed a short time ago by laparotomy.
The youthful lady in question, desirous of marrying and still
menstruating, need not abandon all hope of ever giving birth to
children.

In a case of hydrosalpinx the diagnosis is often quite easy,
its shape being the same as that in pyosalpinx. The pains,
however, are much less severe, a factor which assists us in defin-
ing the tumor more sharply. Colicky pains occur seldom or
never. Cases of double-sided hydrosalpinx may run their course,
as I know from experience, without any trouble whatever. One
patient complained only of a discharge. At the international
medical congress held in 1892 in Berlin, L. Landau spoke of a
certain sensation of undulatory movement as being characteristic in hydrosalpinx. This impression made on the exploring
fingers might be owing to the thin-fluid contents of the tumor and
its slighter sensitiveness.

In regard to the causes of hydrosalpinx I have since gathered considerable experience, all tending to confirm the statement of Landau, that this disease in a majority of cases is the
consequence of gynecological treatment. In rare instances it
may also develop from hæmatosalpinx—the blood being absorbed
(hence hæmatosalpinx, if a correct diagnosis were possible,

does not indicate massage treatment), but in rare cases replaced by water.

In a majority of cases hydrosalpinx arises from inflammations of the tube, caused by chemical irritations, as wiping out of, or injection into the cavum uteri. According to *Schultze* such inflammations may also result from irrigations of the uterus. However, they are not due to an insufficient backflow or high pressure of the fluid, but most likely to an antiperistaltic action of the uterus and tubes.

It thus happens that during irrigations of the uterus the patient will complain of nausea and pain in the side, however well the irrigating fluid may return; vagus pulse sets in (often only forty-eight beats), cold perspiration, contracting pains, which latter yield only slowly to morphine treatment. These so-called uterine colics occur more frequently and more intensely after injections of iodine, and after cauterizations of any kind, except in cases in which only the cervix is wiped out (Kaltenbach).

If we make an examination in such cases we seldom fail to find the tubes in a sensitive state; often also considerably extended by fluid. Under such circumstances it is well to interrupt the treatment of endometritis as soon as feasible, and to see that the tubes become emptied and healed, in order that the isthmus might become obliterated. To be sure, such an inflammation, due to chemical irritation or cauterization, runs its course without high fever, and hydrosalpinx occurs only when the fluid extends to the pelvic peritoneum. The ostium in such a case might easily become agglutinated and give rise to hydrosalpinx. The latter may, perhaps, also develop from pyosalpinx by absorption of pus and substitution of the same by serum.

I once treated a case of globular cyst in the tube, having a diameter of seven cm., in which, it was alleged, abortion had taken place four weeks previously. Before I had made a definite

diagnosis, however, I noticed that the cyst became smaller, and
hopefully I continued the massage. Gradually the contents
became diffused, the windings of the tube became palpable and
at last it assumed a slender and unchanged form.

In the beginning I thought I had to deal with hæmatosalpinx, but later on I became doubtful about it. A real ovum had
not been observed, but merely shreds. Menstruation had twice
failed to take place. Subsequently I believed it to be a case of
tubal pregnancy, and in this opinion I was corroborated by J.
Veit's vivid presentation of the conditions as found in such cases
given by him during the meeting of the gynecological congress in
Freiburg. I declared that massage, after previous destruction
of the embryo, was the best treatment, because it alone would
insure a cure with preservation of the tube. This statement
created a sort of consternation among some gynecologists, but I
still believe this method to be entirely unattended with danger.
Unfortunately I have not met with another case of this kind to
enable me to make further tests. The embryo was dead, otherwise the oval sac would have grown larger instead of smaller.
Massage will hasten and complete the absorption of the individual parts of the ovum, which sometimes may, perhaps, be brought
by tubal abortion into the abdominal cavity or into the cavum
uteri. *Von Winckel* collected a number of cases in which he
destroyed the ovum by morphine injections, and in which the
absorption was accomplished quickly and without danger up to
small remnants of about three to four cm. in diameter. Without
doubt, it would have been an easy matter to disperse even these
remnants and leave the tube in a healthy condition. Should
the absorption, however, take place rapidly, a hastening of the
same would only do harm.

One case in which Von Winckel used the injection after the
sixteenth week of pregnancy, perished owing to absorption of

decomposed products of the ovum; no doubt, it would have been too late to use bloodless methods in this instance. In case of a large ovum, beyond the twelfth week, I would advise waiting after the injection, until the oval sac has evidently ceased to grow smaller, and then to disperse the rest by massage.

CHRONIC PERITONITIS.

Chronic inflammation of the pelvic peritoneum is distinguished in especial as chronic parametritis, periovaritis, perisalpingitis, pericystitis and periproctitis.   If the inflammation extends over the whole pelvic, I call it according to *Heitzmann*, pelveoperitonitis, not as Russian authors do, perimetritis, in a broader sense.   Infectious germs reaching the peritoneal cavity by way of the Fallopian tubes are not necessarily always the etiological factors of such inflammations, although I have seen cases in the autopsy room in which every cross section of the tubes emitted a drop of pus on pressure, but where ascending purulent endometritis of the uterus had been the cause of pelveoperitonitis.   On the other hand I have observed instances where the patients complained of sudden severe pain in the abdomen, with vagus pulse, a sense of oppression and also abdominal pain on breathing.   An immediate examination in such cases did not yield any satisfactory explanation for the complaints; but in the course of the following day a characteristic condition of retrouterine hæmatocele could be made out.   Fever appeared only after some days.

Etiologically speaking, in four instances I only learned the patients had made, during their menstrual period, a long, laborious walk with wet feet in fresh-fallen snow.   In another case a heavy rain had drenched the patient to the skin during menstruation.   In another a cold vaginal douche, in still another a wading through cold river water, were the supposed causes which increased the physiological bleeding from the Graafian

follicles to pathological dimensions.   In another series of cases
the same effect seems to have been produced by *Schultze's* treatment of endometritis presumably continued too near the time of
the menstrual period.

These causes of retrouterine hæmatocele are enumerated
here at such length, because I do not regard them any more as
indicative of massage.   In the beginning I treat this disease with
antiphlogistics and narcotics and later on with warm external
applications, warm irrigations and warm enemata, for the purpose of furthering absorption.   In most cases, however, chronic
adhesive peritonitis remains after the absorption of the blood
effusion.   It is this peritonitis which must be removed by massage as early as possible.   When I first undertook to test Brandt's
method I used to massage every hæmatocele.   But I have found
out since that the absorption of the blood effusion is not sufficiently hastened thereby to reward the physician for his trouble
and to compensate for the inconvenience arising both to the
patient and to him.   Absorption is better left to nature, which
will accomplish it in double the time and keep the patient confined to her bed six weeks instead of three.   Moreover, there is
danger, especially in the first days, of starting a fresh hemorrhage.   Fever in itself would contraindicate a hastening of absorption, brought about by increasing the area of absorptive material
through massage.   I do not venture to decide whether the fever
is aseptic, so-called absorption fever, or whether it is caused by
an infection of the blood effusion from the uterus, tube or intestine.   For my part, I am inclined to assume the latter.   In
many cases fever set in together with acceleration of the pulse,
a few days after the effusion of blood.   In several instances the
hemorrhagic effusion had been preceded by antiseptic intrauterine
treatment which undoubtedly left the uterus sterile; in all cases,
however, there was a history of long continued constipation.

The first indication therefore is to avoid a stagnation of faeces by means of early energetic cathartics. In using ice and narcotics we have to discontinue the application of the former as soon as its anodyne effect ceases, ice finally producing pain instead. In such a case we substitute ''heat all around.'' Cathartics, however, should be used either way. But we can diminish to a great extent the painful sensation caused by the radiation of cold into diseased tissues by placing a wet cloth under the ice bag.

If I call the bleeding from the ovary caused by a severe refrigeration during the menstrual period the main etiological factor of hæmatocele, I differ from the opinion of other authors, who consider the rupture of intrauterine pregnancy the principal cause. Among fifty cases coming under my observation during the last years, there was only one in which menstruation did not appear and where a piece of membrane came away which might have been regarded as decidua spuria.

Although there are cases in which the blood effusion is absorbed without leaving any traces or symptoms whatsoever, in a majority of cases pseudomembranes remain, gluing the pelvic organs together, checking their mobility, and disturbing their functions. These pseudomembranes may be the sequelæ of a so-called fibrinous plastic exudate from the inflamed parts of the peritoneum that lie contiguous to each other, and have become irritated by the bloodclot, or they may be the remnants of liquid effusions. In either case it is the first duty of the physician to remove them by rubbing as soon as possible. Of course, it would be still better if we could prevent agglutinations and disperse the exudate before the pseudomembranes had been formed.

We beg to be excused for this short digression upon the etiology and treatment of retrouterine hæmatocele. After all, it is only an apparent departure from the theme, since in the cases under discussion hæmatocele was the cause of peritonitis and therefore the causes of hæmatocele are the mediate causes of peritonitis. I will now discuss the other etiological factors. of pelvic peritonitis.

Not only by direct infection from the tubes or through the medium of hæmatocele, but also by an inflammation of those organs which are covered with peritoneum, may above disease be produced. Every ovaritis is accompanied by more or less periovaritis, and in every case of parametritis the peritoneum, covering the respective parts of the pelvic connective tissue, participates to a greater or lesser extent, etc. The peritoneum of the bladder and intestines resist the longest the inflammatory processes of the respective organs. But if both surfaces of the peritoneum which are in contact with each other become inflamed agglutination of the same will be the result, as I have remarked before. There are, moreover, chemical, thermic, and mechanical irritations which may produce such inflammations, not to speak of traumatic influences prevailing mainly during operations. Tincture of iodine, solutions of corrosive sublimate. etc., may by virtue of antiperistaltic action of the tubes, get into the abdominal cavity during intrauterine treatment. Too hot injections, too hot mud baths, etc., are apt to produce a kind of scalding of the peritoneum. Mechanical irritation, however, is the most common cause of inflammation. In nearly every case of retroflexion the uterus rubs against the peritoneum of the posterior pelvic wall, and sooner or later the visceral peritoneum becomes agglutinated to the parietal peritoneum. I can furnish characteristic proofs for every one of these etiological factors.

Pain in the region of the inflamed organs and various functional disturbances are commonly complained of in this kind of disease. In one case I found the peritoneum of the bladder adherent to an intestinal coil. If one of the organs performed its function, that is, shortly before defecation and micturition, the patient experienced very painful tenesmus of the rectum and bladder, which she called "Doppelkraempfe" (double cramps). The diagnosis could only be made after the bladder had been explored and carefully palpated. The intestine was transversely adherent to the apex of the bladder. It was loosened bimanually in three sittings, in the same manner as we loosen a retroflexed uterus according to *Schultze's* method. Massage gradually improved the condition to such an extent that both organs performed their functions painlessly and regularly, and that tenesmus occurred only in exceptionally severe functional disturbances of one or the other organ.

In case the fimbriæ become agglutinated, thereby causing occlusion of the ostium fimbriatum tubæ, or when the tube becomes bent at an acute angle, sterility is a frequent consequence. Strange to say, agglutinations of intestinal coils comparatively seldom interfere with intestinal function. In some instances the peritoneum is intensely sensitive, and an attempt to locate the organs by strong pressure bimanually would cause excruciating pain, sufficient to make the patients cry out. These cases seem to be neuralgic in character, the pain being most likely caused by pressure of contracting pseudo-membranes upon nerve branches, whereby lancinating pains are made to radiate towards the periphery. On the other hand, the constant dragging of the adherent organs while in function seems to produce hyperæsthesia of the peripheril nerve-endings, as in hyperæsthesia of the hymen in case of .vaginismus.

Diagnosis.

Peritonitis, uncomplicated by inflammation of those organs covered with peritoneum, manifests itself by sensitiveness on touch, vagueness of limits of these organs, and absence of essential enlargement of the latter. In making a differential diagnosis between peritonitis and parametritis, we have to consider that, in the former disease, the uterus is not displaced laterally during the stage of swelling, nor dragged towards one side during the period of shrinking. If a palpable effusion is present (in peritonitis), the vaginal vault and overlying tissue are not immovable against the swelling, as it would be in parametritis. Besides, the swelling has not the spherical rotundity characteristic of acute parametritis, the originally liquid effusion not having stagnated at the margin of the uterus, but formed a layer over the same. Of still greater diagnostic importance in larger exudations is the appearance of a tumor behind the vaginal portion of the cervix and the characteristic displacement of the cervix in a forward and upward direction. This tumor is formed in the following manner: The effusion, having entered the abdominal cavity, sinks downward according to the law of gravity, and distends that part of the space of Douglas which normally forms simply a capillary fissure. I mean that part which is bounded above by the folds of Douglas, and the depth of which I found varying from one-half to twelve cm., according to numerous anatomical measurements made by me.* The French authors call it cul-de-sac, but in this country it is not distinguished by any particular name from the part which in case of anteflexion of the uterus is always filled with intestinal coils. The amount ·

---

* In nine cases out of fifty-seven subjected to autopsy the depth of the cul-de-sac could not be measured (pseudomembranes, etc.). The average measurement was 35 cm., the maximum 12 cm., the minimum ½ cm. The depth most frequently found was 2½ cm.

of the forward displacement of the cervix and the genesis of the
retroflexion or anteposition depend upon the greater or lesser
depth of this space.   The above mentioned condition, however,
is found only when a greater effusion (serum, pus, or blood) has
taken place in the abdominal cavity.   In the beginning, when
the blood has not yet coagulated, or pus and serum have not
yet been encapsulated, the fluid escapes the exploring finger and
characteristic results are hardly obtainable on palpation.
Besides, any unusual examination in case pus is present is very
dangerous, as local peritonitis might be converted into a general
one.   In such cases we have to depend upon the thermometer
and be guided by the fever.   A local diagnosis, however, can be
made with more ease and less danger as soon as the intestinal
coils, floating upon the fluid, become agglutinated, which latter
process usually develops in short time.   But should there be the
slightest suspicion of the presence of pus, it is far better to post-
pone the local diagnosis as long as possible and to proceed with
the utmost caution.   In making an examination in case of retrouterine hæmatocele, where the effusion is solely blood, soon
coagulating and facilitating the diagnosis by its resistence and
by the sharply defined contour of the space of Douglas, we will
find the upper boundary very vague and far from forming a level.

Treatment.

The treatment has already been pointed out in general.   It
consists in massage, alternating with attempts to separate the
adherent organs.   Pus contraindicates massage (see above); the
old rule, " Ubi pus, evacua," (where there is pus, evacuate) holds
good here as well as in other cases.   Blood and serum do not

require massage. The only thing to be done therefore is to *pre-vent* the formation of adhesions, provided the patient comes early enough under our care, so that we can determine for ourselves the best time for the beginning of massage, or to break up already formed adhesions, if we have to deal with cases which have run their course. In the strict sense of the word, however, we can but rarely speak of these processes as running a certain course and to be done with, because as long as adhesions of the internal genital organs remain, they will cause pains with but few interruptions up to the climacteric stage, and the adhesions between intestines and bladder at times even during the whole lifetime of the patient. If the old rule, that the longer the inflammation has lasted, the longer the treatment will last, holds good in massage, we can easily understand that this maxim is especially applicable in inflammation of the peritoneum, where soft plastic masses are converted into membranes, the breaking up of which would become a very difficult or almost impossible task.

According to the observations of *Fritsch* the small intestines, because in motion, do not become adherent to each other even after their surfaces have been injured during an operation (Annual Report 1892). Thus after abdominal operations, too, it is only necessary to shift the intestines against each other by slight massage in order to prevent the formation of adhesions. In the fall of 1888 I operated on a large cystomyoma. Numerous adhesions had to be ligated. The operation lasted two hours and a half, and at the time I was still operating under antisepsis with parietal fixation of the stump. On account of constipation and abdominal pain I massaged about three months after the operation, to prevent the formation of adhesions. A few years later I saw the patient again. An abdominal hernia of the size of a man's head had formed, not in consequence of the massage, but on account of the retraction of the stump from the abdominal scar—until now my only laparotomy followed by abdominal hernia. The patient gave her consent to an operation, which enabled me to establish the fact that not a single adhesion had been formed; nor were the intestines agglutinated to each other, nor the intestines to the omentum, nor one or the other adherent to the abdominal wall. Since that time I have advised my patients to knead the field of operation slightly, once every day for a few months after the operation and under avoidance of the abdominal wound, until all pain has ceased.

The sensitiveness of the peritoneum, often lasting for quite a long time, disappears in this way so much quicker, and the patients are guarded against intestinal adhesions with all their dangers. Bimanual massage is both impossible and unnecessary; therefore, the patients may undertake this quite simple procedure by themselves.

In the course of this summer Brandt asked me by letter if I had tried massage after operations, saying that he had obtained splendid results from it. I was very much pleased to have arrived at the same conclusion independently of Brandt. However, if firm adhesions have been formed, our patience will be severely taxed. I, too, must confess that, after trying unsuccessfully to loosen the adhesions in five or six massage-sittings, I resort to anæsthesia and, as I have seen it done by my teacher, *Schultze*, proceed to loosen the fixed ovaries and retroflexed uteri.

It is surprising to notice how even uteri that could not be loosened under anæsthesia, may be freed gradually by massage. *Brandt*, for instance, loosened a uterus by means of massage which *Schultze*, as he told me himself, had repeatedly but vainly tried to free under narcosis. I had a similar experience with an adherent ovary. It may be that the pseudomembranes become softer and more succulent by long continued massage

and repeated attempts at loosening. The menstrual period is
especially apt to produce such results, hence massage treatment
ought not to be interrupted on its account.

In this connection we may aptly refer to the massage treatment of larger myomata. Patients afflicted with such tumors
commonly complain of abdominal pain, and in many cases
pseudomembranous adhesions are present. These morbid conditions may be produced either by the stretching and the irritation of
the peritoneum caused by the growing myoma, or perhaps by the
chafing of the tumor surface against the serous covering of the
intestines. Not infrequently these adhesions become numerous
and very vascular, and if the tumor is growing steadily it may
exert pressure upon these vessels, causing stagnation of the
blood and formation of cavities in the tumor. I am also of the
opinion that under such conditions the climacteric stage does not
prevent these tumors from growing, because they receive ample
nourishment through these vascular adhesions to increase in size
in spite of the fact that the uterus grows more anæmic. Thus
it can be easily seen how reasonable it is *to prevent* the formation of these adhesions, especially in cases approaching the
menopause; and in every case in which there is a sensitiveness
on or about the tumor, we should employ massage and antiphlogistics. As soon as the tumor has ascended beyond the small
pelvis, the patient, if intelligent, may be instructed to use massage herself. For less intelligent persons, especially those coming
from rural districts, we are often compelled to prescribe some
kind of liniment, in order to arouse the patient's confidence in
regard to the " rubbing."*

---

*What is said here of self-massaging in cases of myoma, is applicable also
for the gravid uterus. While the growing organ ascends into the abdominal
cavity, peritonitic and parametric fixations are stretched. In paratyphlitis and
perityphlitis, and, in fact, every inflammation above the pelvic cavity, the
patients may do the massaging themselves, but in cases of inflammation in the
pelvis self-massaging is attended with success only when neither shortening
nor growing together has taken place.

Massage, therefore, is not only palliative in relieving pain, but also exerts a definite curative influence. For the rest, I agree with the statement made by Freudenberg, of Kreuznach, that it is possible to remove myomata entirely by massage, I, myself, having removed a few small ones. Interstitial fibromata are changed into the subperitoneal or subserous variety. In a few subperitoneal fibromata I massaged the pedicle, the nourishment becoming lessened and the small tumors being absorbed. Brandt records the same results, but regards them as exceptions. However, I am sorry to say that these favorable exceptions can be achieved only in small tumors (2 to 3, centimeters in diameter).

PAINTING.

In attempting to remove chronic peritonitis in the cul-de-sac, we encounter many difficulties. It is an impossibility to reach this part of the space of Douglas by means of bimanual palpation. In many cases we are not able to detect any sensitive spot in the pelvis, and yet the patients persistently complain of pain when assuming a sitting posture or during defecation, these pains being aggravated just before or during menstruation. If, during the examination in such cases, we press against the posterior vault of the vagina, we are enabled to locate the source of these pains and, by passing the finger up and down, to detect small irregularities in the tissue lying behind. In these cases the forefinger, or two fingers, make stroking movements, the touch surface being turned towards the back, in about the same manner as a painter would paint or pencil such a cavity with his paint brush, the posterior and lateral pelvic wall meanwhile serving as a support. This kind of massage, done solely with the inside finger, received from Brandt the hardly intelligible name of "painting" (Swedish "malning"). Outside of the above cases it is used only in inflammations in or beside the rectum. Moreover, it is very fatiguing, and one has to press firmly backwards to avoid the anterior parts of the external genitals. Cases in which a supersensitiveness of the peritoneum exists (hyperæthesia peritonei of the old authors) do not always indicate the presence of synechia, which may be entirely absent. Massage does not accomplish any or much good in such cases, but in some I applied electricity with satisfactory results. Using a combination of faradic and galvanic currents of only a few milliamperes, I succeeded in diminishing the hyperæsthesia and in removing any existing pseudomembranes by massage.

STRETCHING.

Although the severing of peritoneal adhesions is often manipulated in the same way as the stretching of a parametric band, the description of the manipulation is not so important as the object to be gained through it. For this reason I thought it a better plan to put off the description of "stretching" until now. In treating chronic peritonitis, we intend to separate the adhesions, whereas in chronic parametritis we stretch them. Although we apply our fingers in the same manner in separating adhesions as we do in stretching them,—which latter method is used chiefly in cicatricial contractions of the pelvic connective tissue,—yet the principle is entirely different. A chronic parametritis which has lasted for some time results in shrinkage of the perivascular connective tissue of the pelvis, the vessels of which are so abundantly distributed. As a first consequence we have here a disturbance of the mobility of the uterus and later a displacement of the same, not to speak of other phenomena, to be discussed later on. These bands may run along the uterine veins, in which case we have to deal with parametritis posterior (Schultze); if they accompany the spermatic veins we have a case of parametritis superior, or parametritis spermatica, as described by the author. If both these vascular regions are inflamed, I speak of it as parametritis totalis. Such cicatricial contractions may also run along the venæ vesicales (paracystitis) and along the obturator veins (parametritis anterior). The last three draw, or better "fix," the cervix towards the front and a retroflexion of the uterus is the consequence. Now let us suppose that we have to deal with left posterior parametritis. In this case a band of about the thickness of a finger extends from the region where the fold of Douglas is attached to the side of the uterus backwards toward the region of the sacro-iliac articulation.

Should this band become contracted, the cervix is strongly displaced to the left of the median line, the anteflexion of the uterus is increased, the point of flexion is in a more or less rigid state, and we have a picture of the pathological anteflexion of the uterus before us.   First we try to decrease the sensitiveness of the parametric band by a few applications of massage, and then we are enabled to begin with the stretching.

We place both fingers of the left hand against the left side of the uterus in the left vaginal vault, and do the same with the finger tips of the right hand from above.   We then push the uterus with both hands to the right as far as the sensitiveness of the parametric band will permit.   This stretching of the band produces renewed pain, which we remove by massage.   By alternate stretching and massage (for the removal of sensitiveness and swelling) we stretch the band, although it may require a great many sittings sometimes, and remove it finally perhaps in its entirety by massage.   One can be satisfied if he succeeds in pressing the uterus, without making any effort or without causing any pain against the opposite pelvic wall (in this case the right one).   In the same way we proceed, ceteris paribus, in a case of right posterior parametritis, except that the fingers are applied to the right side of the uterus in the right vaginal vault and that the uterus is pushed towards the left side.   In superior parametritis the main force should be directed to the region of the cornua, and the fundus uteri should be pushed towards the opposite side and downwards, because the bands running along the spermatic vessels extend towards the region of

the kidneys. However, if the uterus is fixed in front, we apply
the fingers to the anterior vaginal vault and in case of a one-
sided fixation push the cervix towards the opposite side and backwards. If we have before us that frequent form of retroflexion
which is mostly due to fixation of the cervix towards the left
obturator foramen, we must push the cervix towards the right
sacro-iliac articulation and if possible as far as the pelvic wall.

There is another method, which we may follow with advantage. As soon as the patient can bear the stretching we make
the band as tense as possible with one or two fingers in the
vagina, while at the same time we massage the band with the
outside hand.

This combination of massage and stretching, used either
simultaneously or alternately, embodies indeed the peculiar and
essential features of Brandt's method. Both massage* and
stretching** have been employed before Brandt's time, but this
happy combination of both manipulations is the most important
part of Major Brandt's inventions.

His liftings of the uterus, by which he imagines he can produce resistance—movements of the round ligaments, are nothing
more nor less than stretching of those parametric bands which
fix the cervix in front.

Although I have always called special attention to the fact
that I value the stretching very highly, perhaps more so than
Brandt himself—it must not be inferred that I proceed with less
precaution. Menstruation, which makes the bands softer and

---

*Of the many gynecologists who have recommended massage I will men-
tion only Martin (Lehrbuch, first edition), Heitzmann (Pelveoperitonitis),
Prochownick, and others.

**Neugebauer seized the cervix with little hooks and made permanent
traction by means of small chains and weights, running over cylinders. Chroback tried to accomplish the same by means of elastic extension. Schultze and
his pupils did the same with Muzeux's forceps in catarrh treatment. Hegar
and his pupils stretched parametric bands (Kaltenbach).

more elastic, just as in chronic peritonitis, it similarly affects
the adhesions, does not at all contraindicate massage, but the
stretching must be done with extreme caution during this period,
as well as during any uterine hemorrhage from whatever cause.
Fever precludes massage for a long time, just as the swelling
produced by stretching contraindicates the latter, until it is
again removed by massage. The swellings are mostly of an
œdematous nature, and it is surprising to notice how large
tumors disappear under our massage treatment. In case we
have succeeded in removing these swellings in one sitting, it
may be regarded as a good prognostic sign, even if the tumor
originated from a puerperium, or from an intrauterine manipulation. These swellings will still be recognizable in the next sitting and in the following; after which, however, they disappear
forever, not to interfere any more with the further treatment.

If now *Jentzer* and Bourcart* interpret the fact that I lay
so much stress upon stretching, in such a way as to convey the
opinion that I originated a new method, I cannot help but regarding this as a very flattering compliment to me, even if they call
this method dangerous, no harm having ever been done by me in
using it. Not only did I often effect a cure very quickly where
other methods signally failed, but among nearly a thousand
(1,000) cases which I massaged, or which were massaged by my
assistants under my supervision, an exudate appeared only in
two cases. In the first case the uterus was retroflexed and fixed,
and the patient had suffered a short time before from retrouterine
hæmatocele. Twice I removed not only the rest of the exudate,
but also the retroflexion and each time, after from four to six
weeks' well-being, the hæmatocele and retroflexion reappeared
without any ostensible cause. When the same thing occurred a

*Jentzer and Bourcart Gymnastique gynécologique et Traitement manuel
des maladies de l'utérus et de ses annexes. Genève: H. George, 1891. S 11.
6

third time, and I proposed to begin anew, the patient left me.
She was treated somewhere else with rest, antiphlogistics and
massage, and was sent to the country later on, but failed to regain
her health. The second case was that of the wife of a colleague,
on whom pelvic exudates had appeared several times before.
Numerous remnants of the exudate had fixed the retroflexed
uterus to the posterior pelvic wall. After I had loosened the
uterus by massage (without anæsthesia), which task took some
weeks, and during which time there appeared only once for a
short while a swelling in the right parametrium, the patient was
well for several weeks. Presently, however, an introperitoneal
effusion appeared twice spontaneously, accompanied by high
fever at its second reappearance. I believe that in both cases
ovaritis hæmorrhagica existed. Dilated veins and rigid cicatricial tissue are apt to increase and prolong a hemorrhage from
the ruptured graafian follicle, so that a blood effusion is produced, which, if infected from the intestinal tract, is absorbed
under severe peritonitic phenomena. Nature usually helps herself in these cases by surrounding such ovaries with peritonial
adhesions, secluding them in this manner from the other peritoneum, but only with more or less severe nervous complications,
while the pseudomembranes are undergoing cicatricial contractions. In each case we have to find out whether it is a better
plan to loosen the pseudomembranes by massage and to restore,
if possible, the healthy condition of the ovaries, or to have the
patient bear her troubles until the climacteric period puts an end
to them, or to hasten its arrival by castration. It is certain,
however, that neither massage nor stretching caused any of the
relapses in the above cases. Having given a general description
of stretching and its technique, nothing remains to be done but to
prove scientifically the necessity of its application in the several
forms of shrinking inflammation of the pelvic connective tissue.

CHRONIC PARAMETRITIS.

Chronic parametritis is an affection, in the treatment of which the method of Brandt records the most positive results and the most signal of victories. This disease being, with the exception of endometritis, the most common of the internal female genitals, the gynecologist or the physician using Brandt's treatment can do the most good to women afflicted with this ailment. For this reason I do not consider it superfluous to describe the symptoms, diagnosis, pathological anatomy, and treatment of the same in systematic order.

SYMPTOMS.

There are four cardinal symptoms of parametritis: Pain in the affected side, constipation, downward pressure, and vesical tenesmus. In most cases all four symptoms are present, but occasionally one or the other symptom may be absent. The order in which these symptoms are enumerated indicates the respective frequency of their occurrence, the first mentioned being the one least wanting, while the last is the one most frequently absent.

In uncomplicated parametritis, that is, if the ovaries and the tubes are not affected at the same time, the *pain* is felt in nearly all cases most severely on bimanual pressure, when we touch the place where the fold of *Douglas* joins the uterus and where the uterine ganglion is situated according to the description given by the old authors. In exceptional cases no complaint is made of spontaneous pain in the affected side, but only of pain experienced on bimanual pressure. This symptom is very seldom absent, if constipation and downward pressure are present. The pain is increased during menstruation, and disappears after the climacteric stage has been reached.

Constipation, in many instances, is treated customarily by artificial means, and occasionally it is only later on that we become aware of the fact that the patient has been suffering more or less from constipation, because she has been in the habit of using various cathartic remedies for years. To many patients this constipation is a source of great annoyance, while others endure it with the greatest patience. In affections of the left side there is nearly always constipation, while it is rarely found in the right-sided affections. At times we encounter cases in which there has been constipation for a while and under appropriate treatment an improvement or perhaps a cure has been effected.

The *dragging-down sensation* is often erroneously believed by practitioners to be an indication of the sinking of the uterus, and the patient is treated accordingly with pessaries. On closer examination we will find, however, that the uterus is not situated lower down, but on the contrary occupies a position above and behind the spinal line (in consequence of shrinkage of the connective tissues). Therefore the above sensation must not be regarded as the consequence of a deeper situated uterus in view of the presence of a healthy parametrium, but as the symptom of an inflamed pelvic connective tissue. Even under normal physiological conditions the uterus descends during any effort increasing the abdominal pressure and during pregnancy, but as long as the pelvic connective tissue is not in an inflamed state the woman does not become aware of this descent of the uterus. An existing inflammation, however, explains the sensation of sinking of the uterus in spite of the actually higher

situation of the latter; hence also the exacerbation of this sensation at the menstrual period, at which time the chronic inflammation is aggravated by the addition of a physiological hyperæmia.

Vesical tenesmus may occur through nervous or mechanical
influences. Branches from the vesical ganglia communicate
with nerves from the uterine ganglia, and by way of the anastomosis between the vesical and uterine plexus an inflammation
of the uterine ganglia radiates towards the vesical ganglia, irritating thereby the nerves which not only supply the upper part, base
and neck of the bladder, but the ganglia of which send branches to
the vessels and muscular coat of the bladder also. Mechanical
influence producing vesical tenesmus manifests itself in such a
manner that, as a consequence of the backward dragging of the
uterus, the part of the peritoneum which covers the bladder is
shortened, interfering thereby with ample filling and corresponding distension of the bladder. But it seems to me more likely
that vesical tenesmus is due to nervous influences, for the reason
that it is often most pronounced at the beginning of the inflammatory stage, whereas shrinkage and displacement occur at a
later period, and not until then is their deleterious influence
upon surrounding structures and organs noticed. In short.
there is no parallelism between the degree of tenesmus and the
degree of shortening of the pelvic connective tissue, but indeed
between it and the degree of inflammation.

In reality parametritis very seldom occurs alone, but is
accompanied by, and in most cases is a consequence of, chronic
endometritis. These two conditions form the picture of pathalogical anteflexion (see above). The symptoms of chronic
endometritis (better metritis, as the inflammation of the mucous
membrane sooner or later reaches the deeper muscular structures) intermingle with those of parametritis. The question

now is, "Which symptoms are due to endometritis and which
to parametritis?" There are two ways to find this out.

We either examine very carefully those rare cases in which
chronic endometritis occurs without chronic parametritis, or vice
versa; or we observe during the treatment of the mixed diseases
what symptoms remain after removal of the parametritis, to
disappear only when a cure of the endometritis has been effected.
The results gained in one way corroborate those of the other.

This old method of *Schultze* used in massage treatment
has shown that the other symptoms of the pathological anteflexion, not yet mentioned, must be referred to chronic endometritis. By these I mean the discharge and the dysmenorrhœa.
The latter designation, however, is not sufficiently striking. At
times it consists in pain in the abdomen—so-called "cramps"—
at other times in backache, again in both combined. In graver
cases these symptoms appear in the intermenstrual period also.
and are aggravated during menstruation; again, in others they
come on only before and after the menstrual flow, to cease
during menstruation. Migraine, or any nervous headache, sense
of oppression, or real pain in the cardiac region or beneath the
sternum, dyspepsia, rising and choking sensation in the throat
(globus hystericus), are classed among the symptoms of endometritis, respectively, chronic metritis. Temporary amblyopia,
singing in the ears, and partial deafness are still rarer complaints.
These so-called reflex or radiating symptoms disappear often
entirely upon application of massage and stretching of the parametritic band, and some of the other complaints also disappear
occasionally, or are at least improved which is especially the
case with regard to the discharge. But we are not able to
remove them entirely by massage, even if we straighten and
massage the uterus thoroughly every day, in case there is a
rigidity at the point of flexion.

The reason why these symptoms improve may be accounted for by the fact that, in treating the parametritic band we at the same time massage the uterus; furthermore, in removing the parametritic cicatricial contraction we liberate the vessels from these contractions, thereby improving the circulation in the uterus and removing congestion. This removal, for the most part, induces a diminution of the hypersecretion, intrauterine germs endowed with sufficient vital energy to keep up the hypersecretion being seldom present. Should there, however, be such germs, we cannot help but doing harm with massage, by increasing the discharge and keeping up the endometritis. The latter, in turn, would again and again re-kindle the parametritis, so that often it becomes necessary, although more dangerous, to treat the endometritis first before we conclude the treatment of the parametritis. Every intrauterine manipulation, as, for instance, the use of a sound, or especially irrigation, or even cauterization, produces a reaction and swelling in the surrounding tissues of the uterus, although the annexa may be entirely free from disease. This reaction may assume dangerous dimensions if these surrounding tissues have already been invaded by inflammation. I therefore strongly uphold the principle, *not to undertake any intrauterine manipulations before the inflammations beside the uterus have been removed.*

It is to be expected, of course, that endometritis cannot be removed by bimanual massage, the inflammation having its seat inside of the organ, whereas only the outside can be massaged. Nor was it easy to imagine how the anatomical changes of the mucous membrane in endometritis or hypertrophy of the glands could be removed by massage, even if we were inclined to believe in involution of vascular dilatation and softening of the cicatricial tissue in those layers of connective tissue which lie beneath the serosa.

Forsooth we have been taught by experience that in a case of chronic metritis we can do no more nor less than remove the sensitiveness of the upper uterine surface by massage, and that only the external layers become softer. Underneath these layers, however, we can distinguish a hard nucleus surrounding the mucous membrane, and in the more superficial area of the otherwise soft organ we can detect hard nodules. These are most likely out of the reach of veins and lymphatics, remain stationary, and can be distinguished from small fibromata only by the fact that they may disappear under further treatment, whereas fibromata become more distinct.

Chronic endometritis is best treated according to Schultze's treatment of catarrh, the manner and technique of which are supposed to be known. During the first twenty-four hours after the introduction of the laminaria tent, which is kept in position by thoroughly tamponing the fornix of the uterus, the uterus undergoes a kind of massage in consequence of its efforts to expel the tent. The mucous membrane and the underlying cicatricial tissue, being the seat of the inflammation, are pressed against the swelling tent, and after the removal of the tent it can be noticed that the uterus has already become softer. We then stimulate the uterus chemically, thermically, and mechanically by daily irrigations, to undergo contractions lasting for several hours. If the uterus is in a state of erethism, these contractions may increase to colic; in torpid uteri, however, the irrigations are at times insufficient to reduce the pathological consistency and enlargement of the organ to normal proportions, and we are then forced to employ more vigorous stimulation in the form of repeated uterine tampons of gauze. This supplementary treatment, especially recommended by *Chroback*, is necessary only in exceptional cases, and often is not tolerated well by non-torpid uteri. But by means of the irrigation treat-

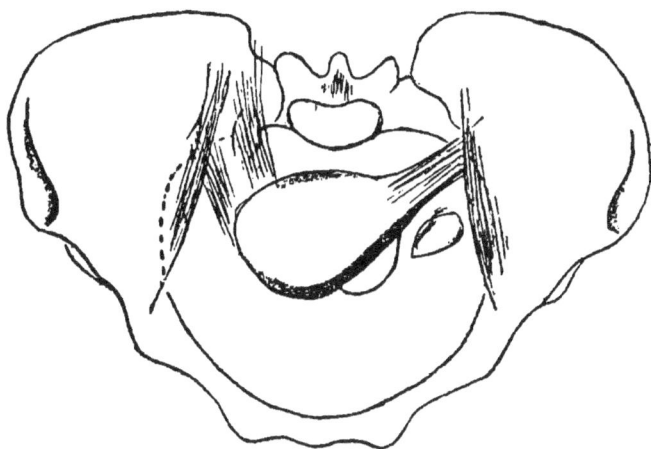

Fig. 6. Parametritis superior dextra and parametritis superior sinistra; the right ovary drawn upwards, ⅓ natural size.

ment of Schultze, with the eventual addition of Chroback's
improvement, we are sometimes able to relieve quite obstinate
cases, in which the uterus is hardened, reducing a cavum of
10 11 cm. length and accessible for the 10-millimeter sound to
a 7½-8 cm. cavum.   If we do not treat chronic endometritis in
such cases, we may soon expect a relapse of parametritis, endometritis having originally been the cause of parametritis.

DIAGNOSIS AND COURSE OF THE DISEASE.

The condition found on bimanual examination has been
repeatedly stated before and consists in sensitiveness of the
cord which fixes the uterus.   The position of the vaginal portion of the cervix is usually extramedian and in most cases
behind the spinal line.   In case of double-sided disease the
cervix may be median, but is always found behind the spinal
line.   We usually find posterior parametritis on the left and
superior parametritis on the right side; we also find frequently
parametritis posterior sinistra and parametritis superior dextra
(see Fig. 6).

The band which has formed along the vasa uterina sinistra
draws the cervix at the point of flexion towards the left; that
which has formed around the vasa spermatica dextra draws the
corpus towards the right side, and, in extreme cases, the uterus
lies obliquely in the pelvis and may at first give the impression
of a lateral flexion.   But on closer examination we find that
the transverse opening of the external os lies nearly parallel to
the sagittal diameter of the pelvis and the anterior surface of
the uterus is turned towards the right side of the pelvis.   In
short, we have to deal with an extreme form of dextrotorsion.
Among nearly one thousand cases that came under my observation during the past few years, I never found one in which
the uterus was bent over its side.   In about thirty cases in

which I found the uterus in an oblique position, it was always
in the condition described above. *Schultze* denies the occurrence of the lateral bending of the uterus, and my experience
confirms his opinion.

W. A. Freund (klinik, Strassburg, 1885) describes the course
of parametritis atrophicans, and says that it can never be cured
entirely, but that it is improved under our treatment, and,
extending gradually all around, spreads over the entire pelvic connective tissue. This description is correct, but it is not
the only type occurring. Of course, from a series of conditions
found on the same patient I could not study and follow the
course of the disease properly, especially as the method
employed was always a successful one, but I had to determine
its course and prognosis from a good many cases by comparing
the conditions as found with the complaints and personal history, as carefully elicited in each case, with regard to the early
beginning of the disease. I so found that the disease in most
cases begins as a posterior parametritis on one side, but that it
usually remains one-sided and only in exceptional cases extends
to the other side.

As soon as the parametric band accompanying the vasa
uterina has once undergone cicatricial contraction, the inflammation will never heal spontaneously. The circulus vitiosus is then
closed. The condensed perivascular connective tissue has a
tendency to constrict the vessels, having about the same effect
upon the blood-flow as the arm-bandage in venesection, that is,
the arterial blood being under a higher pressure enters the uterus
more easily, whereas the venous flow is retarded. The consequence is venous stasis in the uterus, manifesting itself in œdematous swelling of the organ and increased secretion, which in
turn disposes to infection from without and may occasion renewed
endometritis. The latter, in its turn, is apt to produce right-

sided parametritis. The extension of the parametritis from one side to the other is, however, in most cases brought about through the medium of endometritis. Upon the same side the process of inflammation and cicatricial contraction spread by way of anastomosis between the vasa uterina and vasa spermatica, or vice versa, until finally the entire connective tissue on one side of the uterus is invaded. In case the disease remains confined to one side, it will terminate in parametritis totalis and the entire uterus will be drawn towards the affected side. The fundus is not turned towards the unaffected side. If the disease is double-sided from the outset the vaginal portion of the cervix will be drawn backwards and upwards, and the uterus will in a majority of cases remain in a straight position. The shrinking process proceeds in this case from the back towards the front. The best proof that this disease does not heal spontaneously lies in the fact that the patients had often been suffering for a long time, in spite of all seemingly appropriate measures taken by the patients. I have treated cases, or, better, cured cases of this kind, of fifteen years' standing. I, therefore, fully agree with *Freund* (l. c.), who gives a gloomy prognosis in cases in which the patients are left to themselves. When once the parametrium has been invaded by cicatricial contractions the disease cannot be cured by "rest and abstinence from all harmful influences," but it continues, and rest will only favor the shrinking. The latter progresses up to the climacteric period, when inflammation and shrinkage come to an end, with the exception of perhaps a few cases. This relation of the inflammation to the menstrual flow and the reason for the aggravation of symptoms during menstruation, has its foundation in the increase of arterial blood-flow during menstruation, producing augmentation of stasis, increase of œdema and increased emigration of white blood corpuscles. If menstruation ceases, the complaints become less and less and the process gradually comes to a close. The favorable influence of castration may also be accounted for in the same way. But not all the troubles disappear through the menopause or castration, constipation and the so-called nervous reflex symptoms remaining to some extent. If, moreover, we take into consideration the fact that in those cases in which massage and stretching resulted in a perfect cure, not only all these troubles could be removed, but also that these means are less dangerous, it will be readily understood why castration ought never to be undertaken before we have made an earnest effort to remove the disease by above treatment.

I may seem to have deviated somewhat from the subject under consideration, because in describing the symptoms I anticipated a good deal of what, strictly speaking, belongs under the head of treatment. However, my description of the symptoms of parametritis differs from that of other authors; and just as the physiologist has to give a detailed statement of his methods and apparatus before giving a tabulated record of his experiments, so was it necessary for me to mention the therapeutic manipulations by means of which I have gained my experience. Besides, I hope to be able to omit in the description of the treatment what I have already said.

Condensing these experiences into a few lines, we give the following resume: The symptoms of chronic parametritis consist in pain in the affected side, constipation, downward pressure, and vesical tenesmus. The symptoms of chronic endometritis, which in a majority of cases must be regarded as the causative factor, are leucorrhœa, pain in the abdominal and sacral regions, aggravated during exertion and menstruation or present at this time only. In some cases, however, one or the other of these symptoms may be absent. The other symptoms, feeling of oppression in the cardiac region, dyspepsia, globus, clavus,

migraine, amblyopia, etc., are symptoms of irritation of the
sympathetic nervous system, but can be engendered by mere
chronic endometritis, as they may also originate from any disease of the other abdominal organs. For example, some time
ago, in a patient, who had reached the climacteric period and
who had never experienced any abdominal trouble whatever
during the entire course of her sexual life and in whom I established an absolute normality of the pelvic organs, I observed
globus, clavus, hemicrania, and sense of oppression in the cardiac region. The cause of these complaints proved to be catarrh
of the large intestine, and not until this latter was cured did the
above symptoms abate. By the way, I like to remind the reader
of the fact that *Dubois-Reymond* first recognized migraine as
a disease of the sympathetic nervous system, and that, too, on
his own body, and taught us to distinguish between hemicrania
sympathico-spastica and sympathico-paralytica. The cause of
his migraine was a disease of the stomach. Hemicrania is
therefore not exclusively a symptom of diseases of the female
sexual organs. The same may be said of hysterical asthma,
hystero-cataleptic convulsions, hemianæsthesia, etc., in short,
of all those phenomena which are defined as ''grand mal hyste-
rique,'' respectively cerebro-spinal symptoms (Hegar). Sometimes, but by no means always, ovaritis is at the bottom of all
these troubles; frequently we are able to trace the disease to an
inherited neuropathic disposition. To what extent this disposition is inherited is evident from the general symtoms, which
often vary greatly in cases in which the local disease is of equal
intensity.

PATHOLOGICAL ANATOMY.

Schultze has described the clinical condition, the symptoms,
and the course of posterior parametritis. This shrinking parametritis, when occurring on both sides, he recognized as the

undeniable cause of pathological anteflexion of the uterus and
also as the cause of lateroversion, resp. laterotorsion (dextrotorsion and sinistrotorsion), when occurring on one side only.
At the same time and independently of the foregoing, *With.
Alexander Freund* described his pathologic-anatomical discoveries of parametritis atrophicans posterior. Both authors agreed
as to the described parametritis as being the real cause of the
pathological anteflexion, while up to that time the changes in
the wall of the organ itself had been regarded as being the cause
of the anteflexion. *Martin*, for instance, thought he had found
the cause in defective involution of placentation in case the same
was situated at the posterior wall of the uterus. If such were
the case, however, we would expect to find anteflexions only in
women who have borne children. *Rokitansky* (senior, the pathological anatomist), on the other hand, regarded a layer of connective tissue beneath the mucous membrane of the uterus as normal
and called it the supporting framework of the uterus, maintain-
ing that atrophy of this connective tissue at the point of flexion
on the anterior side would produce anteflexion of the uterus.
We know now, however, that the supporting framework is
nothing else but cicatricial tissue which has been formed in consequence of chronic endometritis. *Virchow* has proven that no
atrophy exists at the point of flexion. But such a proof has
become needless nowadays, for the simple reason that all gynecologists believe, in accordance witht *Schultze*, that normally in
a healthy woman with an empty bladder the uterus lies in anteflexion, and not in a position almost corresponding to the pelvic
axis as was formally believed and as some anatomists still
believe. Nor does anybody adhere to the belief at the present
time that a myoma of the anterior wall of the body of the
uterus would produce an anteflexion. Likewise there are only

a few who cling to the antiquated view that flexion of the uterine canal anteriorly would produce any complaints.

While *Schultze* believed the inflammation to be in the
Douglas fold, Freund insisted that it was a hyperplasia of the
more laterally situated vascular connective tissues. *Bandl*[*]
agreed with *Schultze*, put placed more importance upon endometritis as a causative factor than even Schultze himself; whereas
Schroeder and *Fritsch*[**] point to the lack of pathologic-anatomical proofs and interpret the conditions as found by Schultze as
peritonitis of the fold of Douglas.

Thereupon *Schultze* himself indicated the way how to settle
this controversy[***], and I myself, being assistant to Schultze,
undertook in the fall of 1886 the task to solve this question.
The main point was to establish clinically the exact conditions
present in persons apparently destined to die, in order to compare them later directly with the pathologic-anatomical conditions. Although Professor Fuerbringer very kindly placed the
valuable material of the Friedrichshain Hospital in Berlin at my
disposal, yet out of nearly 300 cases whose conditions I had
investigated intra vitam, I was able to make pathologic-anatomical comparisons in eighty-six cases only. But even this small
number was sufficient to enable me to draw some pretty positive
conclusions:

(1) If the uterus was anteflexed during life, it was found to
be so at the autopsy held shortly after death.

(2) If the patient had never suffered from any disease of
the genital organs and did not complain about any abdominal
troubles, the mobility of the uterus was greatest, and it could be
moved to the right and left in the direction of the oblique diam-

---

* Archiv fuer Gynaekologie, XXII. 3.
** Billroth-Luecke's Handbuch, Bd. 1, S. 638.
*** Zwei Gynaekologische Preisfragen, Wiener med. Blaetter, 1880, No.
41 and 42.

eter up to the pelvic wall without destroying the continuity in
the parame·rium.   The fundus uteri could also be lifted in the
median line obliquely forwards and upwards about 4 cm. above
the symphysis pub·s and backwards about 5 cm. above the promontorium.   But if parametritis was present, it could be lifted
only 2·-3 cm.   The mobility was limited to the greatest extent
in retroflexion; and in trying to make the same mobility-test in
the latter disease, the fundus remains 1.5 cm. on an average
behind the symphysis and could be elevated backwards only 0.5
cm. above the promontorium.

(3) Parametritis atrophicans as described by *Freund* is
identical with parametritis posterior of *Schultze*, but, like any
other true inflammation, it does not progress directly along the connective tissue, but along the vessels. Parametritis posterior extends
into the lymph spaces surrounding the uterine arteries and veins.
In recent cases the long-fibered connective tissue has a roseate or
bright red color, and on cutting a thick lymph oozes out, giving
it the appearance of being swollen In cases in which the shrinking process has reached a more advanced stage, it has a somewhat grayish appearance, has lost its long-fibered character, and
become more dense. The adipose tissue, which otherwise lies away
in the vicinity of the pelvic wall, is drawn nearer to the uterus,
and the clusters of fat are harder to separate from the specimen.
The veins are tortuous, often varicose, and contain in some
cases—most likely former puerperal cases—vein stones. The
fold of Douglas is smoothed out in somewhat pronounced cases,
the cicatrizing perivascular connective tissue band is covered
with peritoneum, and only a streak of lighter color marks the
place where the latter ran originally. The band itself cuts like
wet cottonwool. In this place, however, I want to point out
that the folds of Douglas are not of such paramount importance
as one might assume from the experiments of *Kiwisch* and

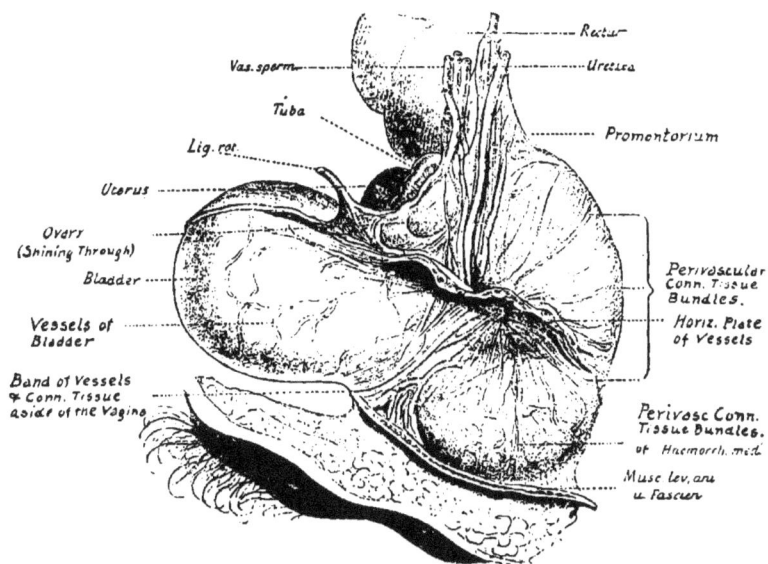

Fig. 7. Pelvic organs loosened up to the levator ani and by blunt dissection showing the pelvic connective tissue with perivascular connective tissue
bands, which, when inflamed and contracted, produce the various dislocations
the inner genitals are shining through. About ½ nat. size.

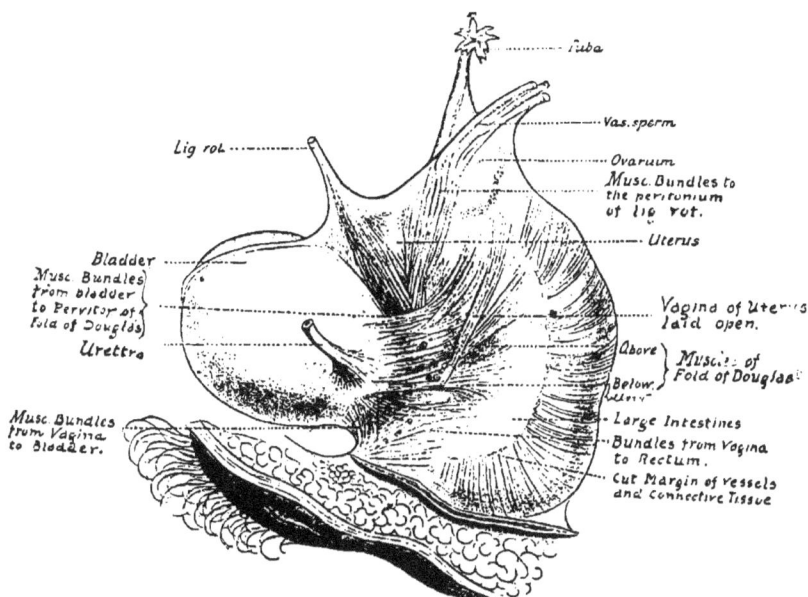

Fig. 8. The pelvic connective tissue is removed in order to show the muscular fasciuli connecting the pelvic organs, especially those which extend directly from the bladder to the peritoneum of the fold of Douglas just as in the male, also those extending to the lig. latum, to the band of the spermatic vessels and to the lig. rotundum. About ½ nat. size.

Tube

Lig ret.

Vas sperm.

Muscles to Pervitor of Bladder.

Ovarium

Uterus

Bladder

Excavativ recto ut.

Fold of Douglas

Urettra

Funnel Shaped space between. Bladder & Uterus

Above } Retr. Uteri

Below } Muscles of fold of Douglas

Musk -
Muscles between. Bladder & Uterus

Douglas

Connective Tissue Space between Rectum and Vagina laid open

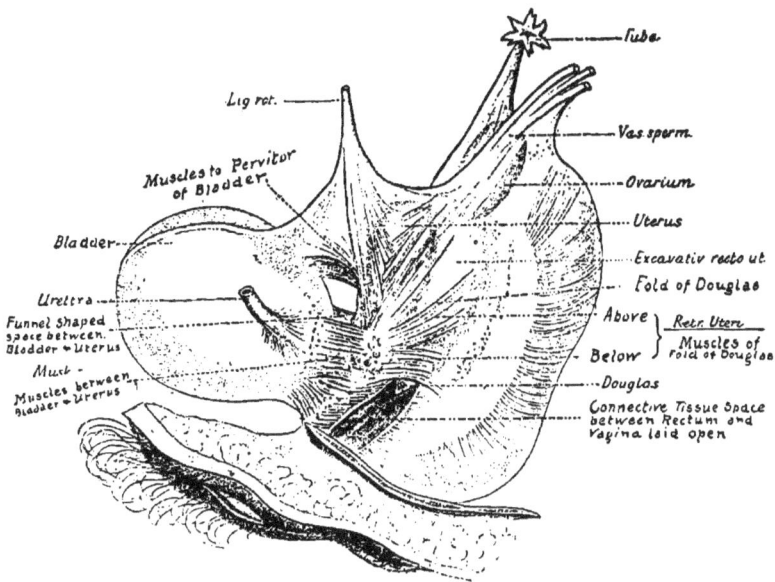

Fig. 9. As in Fig. 7 and 8 the muscular fasciuli partly removed in order to show the deeper ones extending from bladder to uterus and enclosing a funnel-shaped space filled with loose connect tissue, which is laid bare by the removal of the peritoneum of the bladder, also to show those bundles extending from cervix to cul de sac and the fold of Douglas (Retractor uteri Luschka). About ½ nat. size.

Savage, and as has been believed by a majority of gynecologists
since then. The thin fibres of the uterine muscular tissue running out into the peritoneum of the recto-vaginal fold are hardly
one inch long; only that bright streak of connective tissue which
marks the border of the fold runs about the double length. The
folds can be entirely smoothed out by making traction on the
peritoneum, and are not stationary as those peritoneal duplications into which run the ligamentum rotundum, ligamentum
ovari, and the tube (ala vespertilionis). The base of every one
of these three duplications is crossed by dense fibrous tissue and
kept in permanent position. The folds of Douglas, however,
vary in length and height according to the distension of rectum
and bladder, either by giving their peritoneum to that of the
posterior pelvic wall, thereby becoming shorter and more flat-
tened, or by drawing some more peritoneum forwards in case
the uterus sinks forward, and thus getting longer and higher.
In Fig. 7, we have a specimen of healthy female genitals,
obtained by separating the peritoneum from the bony pelvis and
loosening the inner genitals according to *Weigert* by blunt dissection. The different perivascular connective tissue bundles,
alluded to so often, are especially pointed out. Fig. 8 shows
the same specimen after removal of the connective tissue in
order to bring out the smooth muscular bundles uniting the pelvic organs to each other. From the bladder bundles of muscular fibres extend to the folds of Douglas, as in the male. In
Fig. 9 the superficial bundles of muscular fibres have been
removed and the deeper ones are shown, stopping, so to speak,
at the uterus, i. e., in front fibres extend from the bladder to
the uterus and behind from the uterus to the fold of Douglas.
The bundles of muscular fibres connecting the vagina with the
rectum are severed at the side so as to expose a space filled with
long-fibered loose connective tissue lying between them and into

7

which the tip of the cul-de-sac projects. In every one of the three figures, the pelvic organs can be traced through the peritoneum, but the location of the symphysis and promontorium is marked in one figure only.

The peritoneum covering the inflamed parametric band usually shares in the process. Whether the newly formed connective tissue is composed of wandering cells that emigrated from the veins by diapedesis, or whether it is formed from the cells of the fixed connective tissue, is a question that cannot be decided in this special case, as it has not been decided by the pathalogical anatomists in any case of inflammation in any other part of the body.

Parametritis Superior Sive Spermatica.

A picture of the same process as I have described just now and as it presents itself along the vasa uterina I found also along the vasa spermatica. This is described in my inaugural address, pathologic-anatomically.* The vasa spermatica extending towards the region of the kidneys and being accompanied by the lymph vessels, the inflammation naturally pursues the same course. Shrinking of such a band leads first to lateral torsion and, as those inflammations are much more common on the right than on the left side, in most cases to a dextrotorsion. We often find parametritis posterior on the left and parametritis superior on the right side. Such being the case, the condition will, as before mentioned, lead to a dextrotorsion of such a high degree that the uterus lies obliquely in the pelvis (Fig. 6).

*Ueber normale und pathologische Anheftungen der Gebaermutter und ihre Beziehungen zu deren wichtigsten Lageveraenderungen, S. 47; auch Archiv fuer Gynaekologie XXXI. 1.

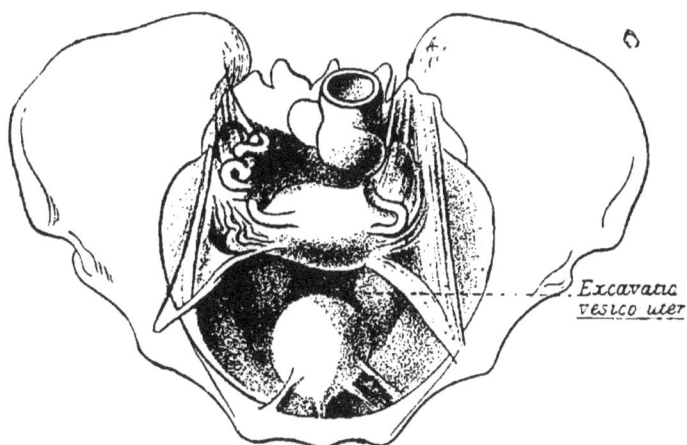

Fig. 10. Double-sided parametritis superior. Ovaries drawn upwards,
empty bladder. ⅓ natural size.

The vasa spermatica supplying not only the corpus uteri,
but also the ovaries and tubes, these organs often share in the
inflammation in the form of adhesive peritonitis.   In such a case
the shrinking causes the ovary to be drawn upwards beyond the
margin of the terminal line towards the region of the kidney.
Eventually the fundus is also drawn up, leading to retroversion
with elevation, as has been described by *Schultze* and *Fritsch*.

(Illustration is taken from case of Mrs. Rusch; the condition
was found during life-time December 4, 1886, and in post-mortem examination December 6, 1886.) This state of retroversion
is similiar to that which is caused by defective descent of the
ovaries and is congenital.

Spiral Turning of the Uterus.

If the uterus lies obliquely in the pelvis, the transverse
diameter of the cervix is parallel to the sagittal plane of the
pelvis, and we have to deal with genuine torsion, i. e., a twisting of the uterus.   Often, however, this transverse diameter of
the cervix is still nearly in the original transverse direction to
the pelvis, only the corpus uteri being twisted, which is possible
only when the uterus has turned on its own axis in a spiral
manner. (See Fig. 6.)

Parametritis Totalis.

When superior and posterior parametritis occur on one and
the same side I call it parametritis totalis, as I said before,
because the whole parametrium of one side is then affected.   In

such cases the region between the internal os and the tubal corner of the uterus is always, but in a slight degree, invaded by
inflammation and shrinking process. We are, however, still
able to recognize these two bands from the internal os and tubal
corner of the uterus in the broad band. A lateral torsion does
not occur in this case, but the fundus and the point of flexion
are drawn towards the same side of the pelvis.

The So-Called Retroversio cum Anteflexione.

Sometimes the inflammation as well as the shrinking process ascends, correspondingly to the vessels of the arch, up to
the uterus, and as a further consequence the point of flexion
will be above the internal os. This condition was described by
Fritsch and others as retroflexio cum anteflexione. *Schultze*
calls it high posterior fixation. I observed a case in 1885 and
1886 at the clinic in Jena (Rosine N.) In this case parametritis
posterior was first established, then high backward fixation, later
an oscillating uterus, i. e., the uterus was now anteflexed, now
retroverted, and finally a retroversion of above described manner formed.

Etiology of Chronic Parametritis.

Schultze and *Bandl* agree that the most common causes of
parametritis are endometritis and in rare cases chronic constipation. *Schultze* also mentions acute puerperal parametritis,
which first displaces the uterus towards the sound side, but
later, during the process of shrinking, draws the uterus towards

the affected side. *Freund* mentions among other causes exhausting loss of strength, as, for instance, masturbation. I should like to include also injuries of the cavum uteri and cauterizations of the same as some of the etiological factors. When I introduced massage-treatment (which, I confess, requires a good deal of time and labor) into my gynecological curative methods, I should have liked very much to have a ready method for the treatment of endometritis, less complicated, less circumstantial, and consuming less time than Schultze's treatment. At that time I used to employ, in conjunction with massage-treatment, all kinds of cauterizations and injections of caustics, even the galvanic current for curtailing the treatment. I always noticed that soft, yielding, œdematous, and very painful swellings appeared in the parametrium. If a parametritic band that had been painless before was present at the time of treatment, it would again become sensitive, whereas, if a sensitive parametritic band had existed, it would increase in thickness and sensitiveness. I accounted for the occurrence of these swellings after the manner of reaction in the vicinity of cauterized wounds on other parts of the body, as, for instance, on the hand. This aseptic inflammation, however, subsides rapidly and is not equivalent to an inflammation caused by the activity of living germs. The latter factor is certainly not present in the uterus. Should there have been previously any germs in the uterus, they are no doubt instantly destroyed by the caustic remedies, all these being exceptionally good antiseptics. But the reaction in the tissues of the uterus and its vicinity, abounding in lymphatic vessels, is much more intense, the collection of lymph in and around the uterus and the emigration of leucocytes much greater here than anywhere else, and in addition to this there are exudations on the free side of the covering peritoneum, producing agglutination of the contiguous surfaces. The effect of these

caustic remedies lies in the coagulation of the albumen, with
necrotization of the surfaces, the caustic itself being chemically
bound down.   Theoretically it might be assumed that, after the
caustic has become completely saturated and has spent its force
as such, germs might settle in or beneath the scab and produce
a genuine inflammation caused by germs.   Even if this objection is only theoretical, the scabs being in reality odorless and
not becoming liquefied, with the exception perhaps of those
caused by mercurials, it is nevertheless a fact that such inflammations in the parametrium produced by caustics do not sub-
side, but linger and produce exactly the same troubles as those
caused by other factors.   Of all the cases that might serve as
proof I will mention only one:

Mrs. F., 22 years old, was treated by me in the summer of
1888 on account of endometritis and left parametritis, left over
from a puerperal period.   The symptoms of the parametritis disappeared rapidly under massage treatment, and when the sensitiveness and shortening of the parametric band were reduced
to a minimum I applied, about the eighth day, first tincture of
iodine and a few days later chloride of zinc and glycerine, both
with a probe wrapped in cotton.   There was an energetic reaction after the first application, increasing after the second.   Certain reasons prevented me from continuing the treatment until a
cure was effected.   The patient coming again under my treatment a year later, I made out an inflammation on both sides.
The previous symptoms had returned with greater intensity, and
new ones had been added.   She also had now a reflex cough, so
violent that the otherwise scanty sputum was at times blood-
streaked.   On this account the patient had been treated for
tuberculosis.   However, I could not detect any demonstrable
physical changes in the lungs, treated the parametritis after
Brandt's method and the endometritis after Schultze.   All symp-

toms, inclusive cough with bloody sputum, disappeared and have not reappeared since. But even more harmless procedures, as, for instance, antiseptic probing of the uterus, and almost always treatment of endometritis according to Schultze, produce like swellings in the parametrium. Experience has taught us that these swellings may lead to permanent disturbances. I am in the habit, after having concluded the treatment of the uterine catarrh, of keeping up massage until all sensitiveness in and around the uterus has ceased. It is necessary, in the majority of cases, and seldom requires more than two days' treatment.

In cases in which caustic remedies have been used, the uterus does not lie in anteflexion, but is held by parametritic bands and peritonitic adhesions on a line with the pelvic axis. The vaginal portion of the cervix had been pulled downwards in order to introduce the Playfair sound, and it had become fixed in this position by the above-described sequelæ. In a majority of cases the external os, as well as the internal os, is constricted by scars due to the caustic, the major portion of which adhered to these places. The greater part of the mucous glands having been destroyed, the discharge is usually diminished, but all the other symptoms have remained. I therefore cauterize only in endometritis of old women and in endometritis gravidarum the lowest segment of the canal. In the former, cauterization does not produce the above mentioned injurious results; in the latter, one cannot use something else. I use tincture of iodine exclusively, as it destroys only the uppermost cells, does not produce any real scabs and is more of an astringent than a real caustic.

With the reader's kind permission, I will briefly relate my experience in regard to the other previously mentioned causes of parametritis. Among nearly one thousand cases of parametritis carefully analyzed by me regarding their etiology, I found only five cases in which endometritis was not present, but in which chronic, congenital, or habitual constipation was supposed to be the cause.   In one case constipation dated from an attack of dysentery, in another case from an attack of typhoid fever.   It is possible, however, that paraproctitis, i. e., an inflammation extending along the hæmorrhoidal veins and due to follicular ulceration, rhagades, etc., in the rectum, has been in some cases the primary cause.   Subsequently the inflammation and swelling may have extended to the anterior trunk of the internal illiac artery and in shrinking may have caused a congestion extending into the peripheral branches, as, for instance, into the uterine veins, producing secondary changes in the uterus and the uterine vessels, which we call endometritis and parametritis.   In such a case it would be difficult to tell whether paraproctitis or endometritis was the primary cause.

Regarding the puerperal origin of chronic parametritis, which origin is disputed by Freund (l. c.), it is too well known that in the course of puerperal endometritis large globular exudations have been found even in the presence of tumor albus cruris.   They disappear, however, after a few antiseptic irrigations of the uterus without leaving any lasting traces.   Such exudations are nothing more nor less than œdemata; and just as in any other infected wound in other parts of the body, for instance, in the hand the swelling after disinfection becomes flaccid and pallid, so also in the puerperal wound of the uterus the œdema, which has spread beyond the limits of the wound, disappears after disinfection.   The nature of these exudates, first described by Virchow as jelly-like and colloid, does not by any means preclude this supposition.   Every inflammatory œdema, if examined on the corpse, will show the same peculiarity.   If, by way of experiment, we kick or beat an animal or inject irritating fluids, as chloroform, into the tissues, the irritated spot, after we have killed the animal and allowed it to grow cold, shows

the same gelatinous material. In a majority of cases, however, not all traces of the puerperal inflammation disappear, but
the acute inflammation merges into a chronic one with all its
characteristics heretofore mentioned. The most extensive parametric bands, accompanied by the severest local and general
symptoms, can in most instances be traced back to a puerperal
period.

Another etiological factor of chronic parametritis, which
has not been mentioned as yet, are operations in and on the
uterus. I know of a number of cases in which inflammation
appeared after a discission. One need not assume that the cut
went through the thin wall of the perhaps still infantile uterus
and exposed the pelvic connective tissue. This is contradictory
to the fact, that chronic inflammation followed in a number of
cases a simple curettement. No antiseptic after treatment was
used in both these cases. The operation wound in these cases is
very similar to the wound produced by expulsion of the decidua
during birth, i. e., the mucous membrane of the uterus.

Again, it is strange to notice that injuries inflicted on the
vaginal portion of the cervix and vagina, even if they do not
heal aseptically, never lead to *chronic* parametritis, resp. para-
colpitis. The cicatricial bands, which, in consequence of such
infected injuries, are formed in the pelvic connective tissue,
become insensitive after a short time. In puerperal lacerations
of the cervix, even if they extend ever so far into the vaginal
vault and expose the pelvic connective tissue, we may expect to
find in spite of infection and acute inflammation a more or less
extensive cicatricial band, which becomes insensitive after some
time, but never do we find chronic parametritis of a progressive
character, as it follows injuries situated higher up. Such laceration ectropia claim our consideration in regard to parametritis
only in so far as they favor the genesis of endrometritis. The

enlarged opening of the uterus and the projection of the cervical mucous membrane into the vagina give ample opportunity to the manifold germs in the vagina to infect the uterus or to irritate the mucous membrane with their chemical products. Endometritis then in turn generates parametritis. It is evident, therefore, that a removal of the laceration does not influence endometritis and parametritis, or, if any, only in so far that the cervix becomes more anæmic in consequence of the cutting and obliteration of a number of its blood vessels, diminishing thereby the discharge. Hence the operation produces the same effect as circumcision of a chronic ulcer, an effect that is produced by any and every operation on the cervix, from simple scarification to the ingenious cone-shaped excision.

This progressive course of parametritis, a course that starts from inflammations situated higher up in the uterus—contrary to those which start from the vaginal portion of the cervix—combined with the fact that after the commencement of the climacteric period or after castration this progressive character disappears and the inflammation of the pelvic connective tissue is checked, is another proof that menstruation with its monthly hpyeræmia conditions the progressive course. *H. W. Freund* (l. c.) has microscopically demonstrated the degeneration of the uterine ganglia. But since it is not settled as yet which nerves occasion the menstrual flow, and not yet explained how menstruation may continue for years after removal of both ovaries, we cannot attach any great importance to it.

Treatment of Chronic Parametritis.

The treatment of chronic parametritis consists in massage and stretching of the contracting band and in removal of the

causative endometritis. The other etiological factors of parametritis are so rarely encountered that they do not deserve
especial mention. According to old therapeutic rules, the first
and most important indication is the indicatio causalis, which is
"Tolle causam" (remove the cause). The second command is
the indicatio morbi. The third indication, the indicatio pallia-
tiva, is an acknowledgment of the weakness of medical science
(therapeutics). The conscientious physician is not allowed to
let the disease take its own course and to confine himself to the
alleviation of the ailments, except in incurable diseases, in
which the first two indications cannot be fulfilled, either for the
reason that cause and character of the disease are not sufficiently known, or because our therapeutic methods are inade-
quate for their removal. Now this latter condition was prev-
alent in regard to chronic parametritis before Brandt's method
became known. But as soon as we learn of methods that will
fulfill the first two indications, it is our most sacred duty to
acquire a knowledge of these methods and of their application.
But those who, priding themselves on their reputation, cling to
palliative remedies on account of their greater convenience,
although convinced of the superiority of these methods, ought
to be designated as unprincipled charlatans.

Remembering, however, what has been said under the head
of etiology, the succession of fulfillment of these indications must
be a different one. We have to remove first the disease and
then the cause, first the parametritis and then the endometritis.
Considering the circulus vitiosus existing between endometritis
and parametritis—since parametritis always sustains stasis and
hypersecretion in the uterus, and endometritis creates every
month an exacerbation of parametritis—the removal of parametritis is at the same time a cure for endometritis. That is to
say, leucorrhœa decreases with few exceptions, more and more

every day, the longer and softer the parametritic band becomes; the uterus itself assuming at first, with the exception of a centrally located hard nucleus, a flabby and rugged appearance, becomes finally very soft. That massage and stretching of the parametritic band diminish the discharge and dysmenorrhœa, has been mentioned before, also that through this manipulation alone the anatomical changes in the uterus, produced by endometritis, are not removed. At the same time, however, we may find out that removal of the endometritis without previous treatment of the parametritis is very hard to accomplish. The difference in the treatment of endometritis in former times as compared with the present treatment, removing the parametritis first, is striking. In this way one therapy—even if somewhat trying—shortens at least the other, since it is not expedient to use massage and endometritis treatment at the same time.

There are only a few cases in which during the treatment of parametritis small subacute relapses of the disease (increased sensitiveness and swelling) occur again and again. Nor does the discharge decrease during the treatment of such cases, but, on the contrary, increases. In these cases there are undoubtedly germs present in the uterus. Bacteriological investigations are as yet wanting. In such cases I do not adhere to the above mentioned injunction, taking up by way of exception and with the greatest care, the treatment of endometritis even before the sensitiveness in the parametrium has entirely ceased.

The treatment is, however, not without danger under such circumstances.

It happens in only a few cases that the parametritic band becomes so firm and large that it cannot be removed even by long continued massage, nor can mobility be restored to such an extent as to enable us to push the uterus up to the pelvic wall of the unaffected side. Inappropriate treatment, as operations in and on the uterus, cauterizations, tents, etc., during flourishing inflammations in the pelvis, are usually the factors that bring about such a condition. These cases, however, with but few exceptions, may be improved to such an extent that the remaining troubles are only trifling ones. They will disappear entirely as soon as the knowledge of massage-treatment becomes general among physicians, so that they may come under appropriate treatment before such bands have through shrinking processes assumed a cartilaginous hardness.

Relapses, too, may occur. By what therapy can such be entirely avoided anyhow? They occur, however, more seldom than in any other treatment, and only a few sittings suffice to remove the relapse provided it is treated early enough. Two or three sittings of ten minutes' duration each are often sufficient. We may say in general that the longer the disease has lasted, the more time it will require to remove the parametritis. The shortest period of treatment which I saw was that undergone by a young woman, who three months previously, in childbed, had acquired a typical parametritis with all its characteristic symptoms. After three sittings the woman was well.

About the technique see pages 44–48, Fig. 2.

It is self-evident that, in stretching a left parametritis, we place the fingers of both hands against the left side of the uterus and push the same towards the right; and vice versa, to remove a shrinking on the right side, we place the fingers against the right of the uterus and push it toward the left side. It does not make any difference in regard to this manipulation, whether we have to deal with parametritis superior or posterior.

In case the ovary has been dragged upwards beyond the terminal line by parametritis superior, it will be drawn down again into the pelvis, and if diseased be made accessible to diagnosis and treatment. Superior parametritis occurs in a majority of

cases on the right side and, like posterior parametritis on this
side, rarely produces constipation; but the inflamed vascular
band extends, together with the ovary, beyond the ilio-psoas
muscle. This accounts for disturbances in walking, because this
muscle, contracting during walking, presses upon the ovary and
nerves of the band.

Competitive Methods.

It is a settled fact that acute puerperal parametritis disappears after removal of the endometritis without leaving any
traces. The same phenomenon we observe in subacute non-
puerperal parametritis, due either to spontaneous cessation of the
endometritis or in consequence of the treatment. But if a
cicatricial contraction has once taken place in the vascular connective tissue band, the circulus vitiosus is closed and the parametritis will not disappear any more by itself. But even in
such cases I have had good results at times with irrigations of
the uterus, for the simple reason that I had to draw the uterus,
which was fixed to one side, towards the middle and downwards,
before I could introduce the tube, by which manipulation the
fixating band would be unintentionally stretched. But such
a result occurs very seldom, because (see above) the intrauterine treatment by itself will again produce parametritis. I
nevertheless stretch occasionally the insufficienty stretched band
completely by means of a hook-forceps applied to the vaginal
portion of the cervix during the treatment of the endometritis, as the ensuing reaction will be removed every time at the
conclusion of the massage treatment.

Rest and warmth all around may be recommended as a
further treatment. This therapeutical regime can be made either

very cheap or very expensive according to the fancy and means of the patient. The cheapest form comprises warm, slightly antiseptic irrigations, warm cataplasms around the abdomen, and warm enemata. By adding a handful of salt to the water we intensify the effect. The latter probably consists only in a production of active hyperæmia and consequently local increase of tissue change. The principle is the same if we add brine, carbonate of soda, etc., or if we order sitzbaths or full baths, with or without the use of the bath speculum. The most expensive but most effectful means of this kind are the mud baths.

All these means, or therapeutic procedures, fulfill the indicatio morbi in effusions. But where there is nothing more to be absorbed, as in a parametritis in which shrinking processes have set in, they have only a symptomatic importance, pain being soothed, even if only temporarily.

A treatment of endometritis having a definite purpose in view, is at least an etiological treatment, even if the "cessante causa cessat effectus" does not, by way of exception, hold good in parametritis that is in a state of shrinking. But those who think they are able to cure endometritis with vaginal irrigations, glycerine-tampons, etc., labor under the same misapprehension as those who try to remove diseases of the throat by means of irrigations of the mouth (Schultze). The tampon, however, on account of its mechanical effect, is of higher therapeutic value than the procedures mentioned under "warmth all around." The hard-rubber ball of *Bozeman* is the same in principle as the tampon. The introduction and the extraction of a tampon resemble, to some extent, a short course of massage and stretching, repeated by every pressure of the abdominal wall and by every change in position, the diseased part being thereby rubbed gently against the tampon. But the tampon has the disadvantage of stretching the healthy side unnecessarily, and the affected side insufficiently, and therefore less perfectly, in a majority of cases, than manual massage and stretching. The same applies to the more elastic tampons made of sheep wool, used in America and England. It matters little in what medicament the tampon has been soaked, as soon as there is nothing left to be absorbed from an effusion. The property of glycerine to abstract water from the tissues signifies nothing else than a local stimulation of tissue change and consequent promotion of absorption. The best results might be expected from potassium iodide, because when used in concentration (potass. iodide and glycerine in the proportion of 1 to 3), as Schultze employs it, we can notice by slight symptoms of intoxication that it has been absorbed. But the parametritic band once completely formed becomes only firmer through the use of potassium iodide. The same opinion as enunciated now in regard to the tampon I expressed four years ago. It was at the time when the ichthyol treatment had reached the high-water mark. In spite of the enthusiastic eulogies of some authors, I clung to the above statements, fully convinced *that mechanical disturbances could be cured only by mechanical therapeutic methods. A chemical remedial substance that removes shrinking bands and elongates shortenings does not exist, nor will it ever exist.* The exact control-tests made at Chroback's clinic, where the results of ichthyol treatment on patients were compared with those of expectant treatment, proved that the remedy does not materially shorten the natural course. In all cases in which ichthyol might promise good results, as in cases in which the parametrium has not undergone any process of shortening, massage is to be preferred, as it cures in but few sittings. This was shown by the control-tests made with this remedy by myself.

The tampon treatment is, after all, the most successful and most rational of all these mentioned, especially when it is pushed up as high as possible. A number of physicians regard the sensation of downward pressure as a symptom of sinking of the uterus, and, although the cervix is in most cases *behind but never* in front of, the spinal line, they push the tampon high up, in order to remedy the supposed sinking. They may remove in this way a parametritis which has not yet advanced to a chronic state, the sensation of sinking is abolished, which latter circumstance confirms the physician in his opinion.

In conclusion the treatment with pessaries must be mentioned. Usually, but by no means always, shrinking of the parametritis produces an increase, mathematically speaking a decrease, of the anteflexion-angle; and chronic endometritis gradually extends to the deeper tissue layers of the uterus, leads to chronic metritis, and produces a rigidity of the uterus at the angle of flexion, i. e., the uterus becomes angle-rigid (winkelsteif). The two symptoms of this rigid flexion, dysmenorrhœa and sterility, were accounted for as being due to the fact that the cervical canal was bent at too sharp an angle—the menstrual blood could find no outlet of the uterus and the spermatozoa no inlet. In consequence thereof the uterus was treated with pessaries and tents, which were expected to straighten it. The fact that in some cases the downward pressure became less or disappeared entirely, again strengthened the gynecologists in their erroneous opinion. The effect was simply that on account of the pessary, the volume of the uterus was enlarged so much that it could not, or at most only a little, descend in case the abdominal muscles were brought into play during any effort, and consequently the downward pressure, i. e., the perception of this otherwise physiological descent with its pains was wanting. But after it has been ascertained that these symptoms are most

8

pronounced in anteversion, a condition in which there is no
angle of flexion, simply because the intense inflammation of the
uterine wall—the metritic rigidity—has leveled the angle, that,
furthermore, the same symptoms may be encountered in cases
in which, owing to a chronic swelling, the cervical canal has
become so large that a probe of 6 to 10 mm. thickness can easily
be passed up to the corpus, it ought to be time to cease torment-
ing the women with a treatment that was based upon erroneous
views and that can have only a palliative effect in single cases.

A tampon or a simple Meyer's ring has the same effect as
an ingenious anteflexion-pessary.

The reason why I dwelled at length upon these other
methods of treatment, is that one cannot dispense with them
entirely in spite of their inferior usefulness. Suppose, for
instance, a lady wishing to consult us calls upon us while enroute
for some point, or has come from a great distance to stay only
one day. Of course, in such a case we cannot undertake massage-treatment. All we can do is to advise massage-treatment
and to select this or that from the other therapeutic measures,
whichsoever might be the most appropriate for the case. The
same has to be done in cases in which massage is contraindicated.

Simplified Methods.

While Brandt himself regrets very much that nearly all physicians use the local treatment only, omitting the hygienic-gym-
nastic treatment in general as requiring too much time and not
being essential for a successful result, others have even tried to
simplify the local treatment. Weissenberg*, for instance, used,

* Centralblatt fuer Gynaekologie, 1889, No. 22.

instead of the fingers of his left hand, an obturator covered with hard rubber for introduction into the vagina and over it massaged the diseased parts. *Saenger* seized the vaginal portion of the cervix with a tenaculum and, while stretching the parametritic band in this way, massaged it with the outside hand (Stretch massage Germ. Zugmassage). Both methods are not only unpractical, but in some cases even dangerous. The finger in the vagina lifts the parts that have to be massaged, but at the same time has to diagnosticate what parts are inflamed, fixed, swollen, etc. Besides we must massage first, until the pains disappear, and then stretch. But if we want to use Saenger's method we cannot massage without stretching at the same time. Suppose we have to massage a conglutination-tumor, in which a purulent salpingitis or perhaps a small pyosalpinx is lurking· what then?

RETRODEVIATIONS OF THE UTERUS.

If in case of retrodeviation of the uterus the corpus and cervix form an angle, we call the deviation retroflexion; if they do not form an angle, retroversion. It was formerly supposed that version—the inclination—was the less serious, and flexion —the bending—the graver disease. But it has been shown that the very opposite is the case.

Retroversion arises when the uterus has rigid metritic walls and the displacement is accompanied by graver symptoms of chronic metritis than when the walls of the uterus are still soft enough to form an angle. Hence versions and flexions do not differ in principle, and they may therefore be discussed jointly. For the same reason it is indifferent how large or how small the angle is which is formed by the corpus and cervix. We may take it for granted even that the sharper, or mathematically speaking, the smaller the angle is, the healthier and more flexible the uterus ought to be. It is therefore idle play to distinguish three degrees of retroflexion.

The third degree represents a milder disease than the first, and the boundaries of these degrees are drawn differently by different authors, so that it does not even facilitate the mutual understanding of the prevailing condition when we speak of degrees. Just as in anteflexion, so also in retroflexion, the flexion of the cavum is without any scientific or practical significance. We did not recognize the pathological anteflexion as a single disease, but had to divide it into endometritis and parametritis, and this principle applies still more cogently to retrodeviations.

It was about the same with hydrops, which was originally treated as a single uniform disease. Gradually, however, with advancing recognition of its causes, a whole series of diseases sprang from it—nephritis, hepatitis, heart disease, hydræmia, marasmus and, finally, tuberculosis of the peritoneum, with often diametrically opposed treatment. But while we have to excuse the empirics, who, by way of experiment, tried a number of remedies on hydropic patients with good results, because they did not know the real causes, we cannot shield our modern gynecologists in the same way, for to-day we know that retroflexion is produced by different causes, and yet, in spite of this fact, they proceed in the same manner. Big operations are performed, more dangerous than the potions of the empirics. At times one may hear or read the remark, " It is generally acknowledged that retroflexion ought to be treated."

At the conclusion the woman is assured that the uterus is now in the right position, and that the symptoms ought to disappear. The woman who in the end does not care whether the uterus lies backwards or in front any more than for the loss of an incisor tooth, as the uterus cannot be seen by anyone, provided only she is freed from pain, finally surrenders herself to fate with all her troubles, or, at least, a greater part of them. Otherwise people might think her to be a peevish, fretful, lazy woman.

SYMPTOMS.

Retrodeviation in itself does not produce any symptoms whatever, but the symptoms ascribed to it are those of the accompanying diseases.

The most common symptoms are those of chronic endometritis and parametritis, which have been discussed before, together with more or less decided reflex symptoms.

For a short time I assumed one symptom as belonging to

retrodeviation, namely, a pain across the sacral region, nearly
corresponding to the brim of the pelvis.   But even this symptom was wanting in a great many cases and did not appear at
the moment when retroflexion recurred, but only in case the
woman exerted herself.   I then found out that it was due to an
irritation of the peritoneum, and occurred only at the time when
or at the place where the fundus rubbed against the posterior
wall of the pelvis.   A peculiar case, in which sensation of
oppression and asthmatic attacks became the two conspicuous
symptoms shortly after the pessary had been removed, may be
accounted for by the same cause.

Subjoined we give the reasons corroborating our proposition as enunciated at the beginning of this chapter:

1.   A great many cases of retroflexion are found quite
accidentally, without the patient having ever hinted at it by a
single complaint.

2.   By removing chronic parametritis, periovaritis, or other
pelveoperitonitis through massage, or massage and stretching,
and by removing chronic endometritis, the symptoms are made
to disappear without retrodeviation being removed.

3.   Relapses of retrodeviation may occur without recurrence
of the original symptoms, but the latter reappear as soon as the
original accompanying diseases recur.

4.   The great number of cases in which the same symptoms
are present in case of anteflected uterus, that is, the great number of so-called pathological anteflexions.

Consequently, and consistently with these propositions, we
cannot hope to be able to remove the symptoms in a case of
retroflexion by forcing the uterus into an anteflexed position.
It would be perfectly correct to remove only those accompanying diseases which cause the patient so much trouble.   But the
displacement of the vaginal portion of the cervix forward, beyond

the opening in the levator ani muscle and the ostium vaginæ favors the genesis of a prolapse. Torsion and dragging of the vessels of the parametrium produce a congestion in the uterus and favor the occurrence of relapses.

It is, therefore, desirable to remove the retroflexion also, and, as every treatment ought to be etiological, we must acquaint ourselves first with the causes of retroflexion.

ETIOLOGY AND ANATOMY.

Retrodeviations caused by tumors of the uterus or its annexa may here be properly passed over. Such tumors may often displace the uterus in every possible direction; why not then backwards? If a few authors, as Schultze, H. W. Freund and myself, have contributed such cases, in order to describe the mechanism of their origin, the displacement nevertheless yielded only upon removal of the tumor. For this reason such cases can have no interest for us in this connection, inasmuch as the point in question is a cure of retrodeviation by means of massage.

Schultze,* besides the case in which a tumor on the anterior wall of the cavum uteri gave rise to retroflexion, has enumerated four causes of retrodeviation:

1. Congenital or senile-atrophic shortening of the anterior vaginal wall.

2. Fixation of the cervix in a forward direction by shrinking inflammatory bands between the vagina and bladder.

3. High posterior fixation of the cervix in case the retroverted uterus is high up.

4. Relaxation of the folds of Douglas and sinking forward of the cervix.

E. Martin mentioned as causes (1) incomplete involution

---

*Schultze Pathologie und Therapie der Lagerveraenderungen der Gebaermutter. Berlin, 1881, P. 124, ff.

of that part of the wall of the uterus where the placenta was situated, which is mostly in the anterior wall, and consequently bending down of the anterior wall of the uterus; and (2) continued recumbent position, especially during childbed.

Fritch,[*] besides those mentioned, gives the following:

1. Fall upon the buttock.

2. Continued overdistension of the bladder during childbed and resulting backward displacement of the fundus.

3. Relaxation of all the organs of the pelvis through masturbation, so that the uterus is at times situated in the front, at times at the back.

Kuestner[**] observed the formation of a retroflexion superinduced by shrinking of a lateral fixation, which occurred in consequence of the distribution of an extensive parametritic exudation, and ventured the opinion that virginal retroflexions were much more frequent than generally supposed and due to imperfect descent of the ovaries.

Von Winckel, in his text-book of women's diseases (second edition, p. 390), classes chronic constipation as an additional cause of retrodeviation, an overcrowded rectum in conjunction with forcible abdominal pressure during defecation pushing the cervix forward, whereby, with the co-operation of a perhaps overfilled bladder, the fundus is bent backwards.

Thure Brandt believes that the fundus sinks backwards in consequence of relaxation of the round ligaments of the uterus.

Another theory concerns retroflexio uteri fixata, erroneously ascribed by Scanzoni to Virchow,—namely, that shrinking peritonitic pseudomembranes drag the fundus backwards and fix it at the same time. Although Virchow[***] himself declared, "On

---

* Fritch, in Billroth-Luecke's Handbuch, Zweite Auflage, Bd. I, S. 703 ff.
** Zeitschrift fuer Gynækologie. Bd. XI, 2.
*** Gesammelte Abhandlungen. S. 828.

the contrary, by the adhesions I simply accounted for the fixation of the fundus uteri," this theory is nevertheless up to the present day regarded as the most plausible, whereas under the most favorable circumstances it could be applied only to the proportionately small number of retroflexions where such fixations occur.

To these fifteen theories I am able to add three of my own: Retrodeviations are caused:

1. By retrouterine hæmatocele, in which the cul-de-sac has extended to a considerable depth.

2. By superior parametritis, in which during the progress of shrinking the band pulls the fundus backwards and upwards.

3. By fixation of the peritoneum of the vesico-uterine excavation to the cervix.

It would be an easy matter to find additional theories concerning the origin of retroflexions, if we perused current literature. Fritsch, for instance, thus classes tumors that develop in the cul-de sac and lift the cervix forward and upward. But I hope that the above will suffice, that nothing essential has been overlooked, and that, when critically sifted, the complicated system may be simplified.

EXPERIMENTS ON DEAD BODIES.

Before entering upon this criticism, I may be permitted to call to mind the results of investigations and experiments made in my post-mortem examinations and published in my inaugural address.*

First of all it appeared to me that, on the part of gynecologists, too, the value of phenomena as observed on corpses was

*Ueber normale und pathologische Anhaftungen der Gebærmutter und ihre Beziehungen zu deren wichtigsten Lageveränderungen. Arch fuer Gynækologie, XXXI, 1.

overestimated.   If the uterus was anteflexed in the living
body, it was thus also in the dead body.   Once I had the opportunity to observe, for three days, a woman with healthy genitals and, apart from the abdominal incision,* uninjured abdominal organs.   I noticed how in the course of time the uterus
was sinking backwards.   It seemed to me as if, during increasing maceration, the intestinal coils behaved physically towards
the uterus more like a thick fluid in which the heavier uterus
becam esubmerged.   Similarly in the living body the much
heavier uterus of the lying-in woman will sink backwards
during continued recumbent position, if intraabdominal pressure and intestinal coils fail to sustain it.   Moreover, it
struck me very much that with the advancing age of the patients
retrodeviations increased.   It is possible that the senile-atrophic
shortening of the vagina has something to do with this phenomenon.   I for my part, I am sorry to say, did not pay any
particular attention to it at the time, but it might be similar in
these cases as in hernia—the longer one lives the more chances
there are to acquire one.   The older a woman is, the more
chances she has had to contract retroflexion from some of the
above named causes.   These facts help us to understand why
the pathological anatomists** have been convinced for a long
time that anteflexion is the normal position of the uterus, while
the anatomists up to the present day have given utterance to
conflicting views.   The former receive their material for purposes of observation, fresh and representing every age.   With
the latter it matters simply whether the relatively few female
corpses generally turned over to dissecting rooms come from
prisons or hospitals, thus comprising mostly elderly people, or
whether there are also youthful subjects among them; whether

* l. c., Plate Ii, Fig. 4.
** Virchow, l. c.

the corpse has been brought from a distance or not; how much time the students will require in laying bare and studying the pectoral and abdominal muscles in order to determine in what condition the situs viscerum will present itself; finally, whether the anatomist in question regards anteflexion or retroversion as the normal position, hence what position would he prefer to find. Let it be stated, incidentally, that a good many anatomists are greatly mistaken in assuming that virgins present unobjectionable material for the study of normal conditions. Even maidens are apt to become deceased. Two sisters, intact virgins, came under my observation, one of whom* had congenital retroversion due to imperfect descent of the ovaries, the other** was afflicted with chronic endometritis, subserous myoma, and right chronic parametritis of an acute type. Unobjectionable material we can obtain only when we are convinced that sexual life has had a typical beginning and progress and that there have never been any symptoms of trouble whatever.

Schultze defines the normal position of the uterus as anteflexio-versio, slight sinistropositio and dextroversio. But the conception of normal position includes such an elasticity of the pelvic connective tissue that the uterus can yield to the movements of the neighboring organs, bladder and rectum. Filling of the bladder induces reposition and elevation; filling of both bladder and rectum, elevation and position of uterus in the pelvis axis. Schrœder did not agree with Schultze in so far as he maintained that filling of the bladder caused retroversion. On suitable corpses I made experiments in such a manner as to fill the bladder with air. I chose air in order to exclude gravitation. The abdomen having been opened, the uterus at first

---

*Ueber normale und pathologische Anhaftungen der Gebærmutter, etc. Arch fuer Gynækologie XXI, Plate VI, Fig. 1–6.

**l. c., Plate V, Fig. 7, 8, 9; also Fig. 13 of this book.

fell backwards, but it was easy to observe that the peritoneum over the bladder was shifting backwards. *Indeed, the perito-neum over the bladder is removable in a high degree and is reg-ularly inserted* AT THE CORPUS UTERI *between its lower and mid-dle third,* NEVER under physiological conditions at the internal os. Owing to this arrangement one may draw the body of the uterus forward by means of the peritoneum of the anterior pelvic wall. Firmly holding now the peritoneum in front, the uterus was placed only in retroposition and elevation, not in retrover-sion. Only in case the bladder is overfilled to such an extent that the elasticity of sound parametric tissues, too, is exhausted can the uterus be retroverted under physiological conditions. The bladder may easily undermine the peritoneum as far as the umbilicus.

Overfilling of the rectum affects the position of the uterus to a less extent, because it gives way to the uterus, and that, too, mostly to the left side. The bladder, which had become voided during the lifetime of the patient, and which, therefore, was still susceptible of contraction before the rigor mortis set in, in which condition it retains the once assumed dimensions even after its voiding, represents a flat-convex disk, whose posterior margin is often situated at a distance of 2–2½ cm. (measured) from the cervix. With the uterus the bladder is connected only by two lateral muscular bundles, the vesico-uterine ligaments (see Fig. 10). The most lateral fibers of this muscular bundle pass by the uterus, reaching, as in the male, the peritoneum of the Douglas space directly (see Fig. 8); the more median fibers, as it were, stop for rest at the cervix— that is, they are inserted there, and other fibers extend thence to the Douglas space, where they are inserted after having run a short distance and form a fold (see Fig. 9). To the extent mentioned above the peritoneum thus lies directly upon the anterior vaginal vault.

Here is an opportunity of its becoming adherent through inflammation, while distension of the bladder leads to elevation of the cervix and to retrodeviation.

The importance of the folds of Douglas, as already stated, has been greatly overestimated. Often I had to bring out these inconstant structures artificially by stretching if I wanted to measure them. At the very spot where these folds, if present, are inserted, the peritoneum is very movable. A traction executed on them is distributed over the whole region as far up as the diaphragm.

Only in the middle, at the promontorium, it is less movable. Upward traction on the peritoneum of the posterior pelvic wall produces elevation of the cervix and anteflexion, respectively increased anteflexion. That one may draw the uterus by the spermatic (ovarian) cords, which, in lean individuals, likewise run out into a special fold of the peritoneum, first laterally, and, by more vigorous traction, together with the ovary, in a backward and upward direction, was a self-evident matter, answering simply the progress of these vessels. Traction on the peritoneum of the anterior pelvic wall induces anteversion and elevation.

After this brief discussion, I shall proceed to consider critically the causes of retrodeviation. Of the four causes advanced by Schultze, three are facts, the results of exact clinical observation, demonstrated pictorially by Schultze himself from illustrations in his periodicals. The fourth cause, however, supported in his book (l. c.) simply by a *schematic* figure, may be viewed as a mere theory, namely, the theory of the relaxation of the folds of Douglas.

THE DOCTRINE OF THE RELAXATION OF THE FOLDS OF
DOUGLAS.

It was this very theory that, at one time, induced me to
draw retroflexion within the circle of my observations, for I
wanted to demonstrate its correctness. For this purpose I
made the experiment described on pages   and   , and only
for the reason that also in perametritis mobility in upward
direction was limited to a great extent, did I include it in my
experiment. The proposition was: If, in reality, relaxation of
the folds of Douglas gives rise to a greater number of retroflexions, this increase of the mobility must be measurable even
in a small number of cases. If this be the case, mobility from
the back towards the front must have been increased, while
from the front backwards, respectively in another direction, it
remained unaltered. However, in nineteen cases, without any
exception, the result was that in retroflexion mobility backwards, as well as forwards and above all upwards, was limited
to a great extent.

The mobility of the uterus was ascertained in the following
way: Having loosened any possible peritoneal fixations, the
fundus was grasped with the fingers and drawn in the front above
the symphysis and thereupon in the back above the promontorium, to such a height that it seemed that the continuity of the
subperitoneal tissues was about to be severed. I found that the
fundus could be moved:

1. Normal anteflexion: Above the symphysis, 4 cm.   ⎰ Average in 20
                       Above the promontorium, 5 cm. ⎱   cases.
   Maximum towards the front, 6 cm.
   Minimum towards the front up to the symphysis, 0 cm.
   Maximum towards the back, 6 cm.
   Minimum towards the back, 1 cm.

2. Fixed anteflexion: Above the symphysis, 2.37 cm.   ⎰ Average in
                      Above the promontorium, 3.04 cm. ⎱ 17 cases.
   Maximum towards the front, 5.5 cm.
   Minimum towards the front, 4 cm. behind the symphysis = —4 cm.

Maximum towards the back, 6.5 cm.
Minimum towards the back, 2 cm. beneath the promont.= 2 cm.
3. Retroflexion: As far as behind the symphysis, 1.5 cm. ⎱ Aver.
      =1.5 cm. ⎰ in 19
      Above the promontorium, 0.5 cm. = +0.5 cm. ⎰ cases.
Maximum towards the front, 4.5 cm. before.
Minimum, 7 cm. behind the symphysis, = 7 cm.
Maximum towards the back, 4 cm. before.
Minimum, 3 cm. beneath the promontorium = 3 cm.

Mobility towards the front being of especial interest, we condense in a brief synoptical table the chief results of our measurements. Measures *above* the symphysis are designated with the plus mark (+); *beneath* it with the *minus* mark (—).

Anteflexion.	Fixed Anteflexion.	Retroflexion.
Mean, +4 cm.	Mean, +2.37 cm.	Mean, —1.5 cm.
Maximum, +6 cm.	Maximum, +5.50 cm.	Maximum, +4.5 cm.
Minimum, 0 cm.	Minimum, 4.00 cm.	Minimum, —7.0 cm.

From this tabulated statement we see very plainly that in nineteen retroflexions there was not a single one in which the mobility reached that of fixed anteflexion, much less that of normal anteflexion. An increase of mobility in any direction, therefore, is entirely out of the question.

In the minimum, as well as in the mean, the limit of mobility by seven centimeters, respectively five and a half centimeters, is indeed quite considerable.

If, in the same way, we collate and compare mobility towards the back:

Anteflexion.	Fixed Anteflexion.	Retroflexion.
Mean, +5 cm.	Mean, +3.04 cm.	Mean, +0.5 cm.
Maximum, +6 cm.	Maximum, +6.50 cm.	Maximum, +4.0 cm.
Minimum, +1 cm.	Minimum, 2.00 cm.	Minimum, 3.0 cm.

We shall find in these figures the limitation of mobility plain indeed, but not quite so striking as in the measurement of the decrease of mobility towards the front. The greatest difference we find in the mean, namely, 4½ cm., and in the minimum 4 cm., so that even from this table we can gather no proofs for the fixation of the cervix towards the front.

The more minute pathologic-anatomical structure of such bands coincides exactly with that of chronic posterior parametritis described before (p. 105). It consists in hyperplasia and contraction of the perivascular connective tissues, tortuosities and varicosities of the veins, and possible presence of vein stones. Contraction of the peritoneum, especially in fixed retroflections in which the pseudomembranes covering the peritoneum shrink, while the uterus lies retrodeviated, may be considered the possible cause. In order to obtain information about additional etiological factors in this way, we ought to group together only movable retroflexions and exclude the fixed ones. Fixed retroflexion of the uterus is due to other causes, as will be demonstrated in a subsequent chapter. But what the above figures plainly show is that limitation of mobility in an upward direction is common to all retroflexions. This limitation becomes evident already in anteflexion, especially if it is determined by mobility in a forward and upward direction. That in the mean, it is less noticeable in backward and upward mobility, but in the maximum exceeds even normal anteflexion, might have been expected from the outset, because the traction axis of the shrinking band lies in the same direction.

Shortly after concluding the above mentioned experiments I became acquainted with Thure Brandt and his method, and I was very much astonished at the remarkable coincidence of his therapy with my results. By means of methodical elevations, so-called liftings, Brandt cures a certain number of retrodeviations and prolapses. My belief in retroflexion as caused by relaxation of the folds of Douglas was shaken as soon as I made a more exact anatomical investigation of these structures, for, on account of their inconstancy, we might have had to deal from the beginning

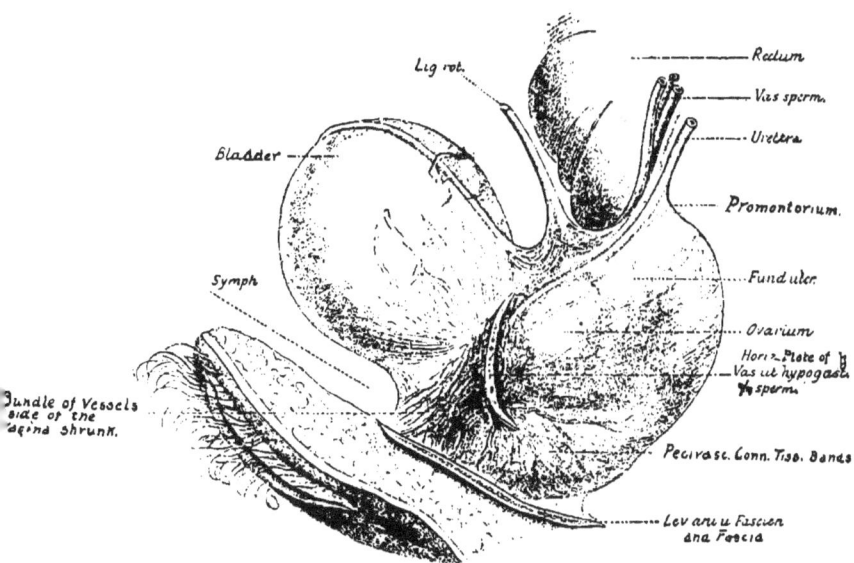

Fig. 11.   Retroflexio uteri produced by left anterior fixation of cervix, the vessels which extend to the side of the uterus are pulled aside. Compare Fig. 7.

with relaxation of the peritoneum of the entire posterior wall of
the pelvis, and consequently the only rational treatment would
have consisted in shortening the elongated peritoneum of the
posterior pelvic wall by operative interference. As soon therefore as I had experimentally tested the etiology of retroflexions,
I desisted from my original plan of experimenting on the dead in
order to find a convenient, harmless way for operative shortening of the posterior fixations of the uterus. If, nevertheless,
such attempts have been made by others, they do not have the
value of an etiological, but merely of a palliative treatment.

If at that time I did not yet dare to eliminate relaxation of
the posterior fixations of the uterus from the list of etiological
factors of retroflexion as completely as I do to-day, the reason
was that, mindful of my past clinical work, I had not infrequently found the uterus quite movable and the posterior fixations quite thin and elastic. The latter circumstance might be
accounted for by the assumption that, together with the cervix,
the denser tissues had been pushed or drawn forwards and thin,
elastic peritoneum drawn down from above. On the other
hand, we had to take into consideration the fact that those
afflicted with retroflexion are not apt to die. It is mostly elderly people or retrodeviations of long standing that constitute the
material for post mortem examinations, while for purposes of
clinical observations such fresh cases are especially received
where retrodeviation produces most troubles, patients not yet
having accustomed themselves to their situation. It is quite
possible that previously there was an increase of mobility which
subsequently became obliterated by consecutive peritoneal and
parametric inflammations. With this object in view I have for
the last six years carefully examined every case of retrodeviation, hundreds of them, and yet I did not find a single one in
which an increase of mobility was noticeable. In those cases
9

that hitherto made me believe that mobility had increased and
the posterior fixations relaxed, there was always a shortening
of the anterior fixations present and demonstrable, provided not
only mobility towards the right and left, but also mobility in a
backward direction was looked into. Never was I able to push
the cervix, especially in the direction of the oblique diameter,
so far back into the hollow of the sacrum that it would lie close
to the posterior wall of the pelvis, as may always be done in a
case of anteflexed uterus with healthy parametria. Often a
band can be plainly followed out running from the cervix towards
the left and anteriorly to the foramen obturatorium sinistrum,
in which case the fundus uteri lies posteriorly to the right of the
promontorium. This is the most common of all types of retroflexio uteri. I was delighted in noticing that my pupils, often
after a few weeks of instruction, were able to diagnose this band
quite accurately, provided they had been instructed at the very
beginning to search for anterior fixations. Frequently such a
band may be lacking and the shortening diffused; in rare cases
the band is on the right side and the fundus lies backwards to
the left side. In reality relaxation at the back is therefore only
apparent, and depends on the fact that the peritoneum of the
posterior pelvic wall is drawn nearer.

In a case like this I was able to observe the formation of
retroflexion and the band which caused it.

Mrs. N., in G., in the summer of 1889, was suffering from
endometritis and left puerperal parametritis, together with
tumor albus cruris sinistri. Upon antiseptic irrigation of the
uterus, the fever and the purulent discharge ceased in short
time, the nearly spherical tumor on the left side disappeared,
and the swelling of the left thigh diminished. In the summer
of 1890 the lady consulted me on account of parametritic complaints. On examination I found the uterus retroflexed, the

cervix fixed by a sensitive band running towards the left side and a second one extending towards the front to the obturator foramen.

The same process which in this case ran an acute course and in which the tumor albus demonstrated how from the simple left parametritis an inflammation descended retrogressively along the congested vasa obturatoria, may likewise occur in chronic cases of parametritis, although less conspicuous.

In another case the condensation or rather shortening of the tissues extending from the cervix towards the front, lay more diffusely in the vesico-vaginal septum or at most immediately beside. The patient had been afflicted with severe cystitis some years ago, which proved to be the cause. In stretching the cicatricial contractions one could feel how the tissues, in moving the cervix backward, snapped in twain with a crackling noise. It was the case of Miss H., of B. Retroflexion was completely cured.

In other cases, however, no such inflammation extending from the cervix towards the anterior circumference of the pelvis could be demonstrated as being the cause of the shortening. Only a short time ago I observed three cases in which during continued recumbent position and distension of the bladder the fundus had fallen backwards. (Only one of the three patients was a puerpera.)

They had been previously under my treatment and daily observation. Inflammation and shrinking towards the front were therefore entirely out of the question. And yet a shortening of the tissues in front had occurred subsequently. The fundus which had fallen backwards displaced the cervix towards the front, and the tissue crowded towards the front had become condensed.

Thus, in many cases where we finally establish a shortening

of the tissues that are inserted in front of the cervix, another
cause may originally have brought about retrodeviation. Certain
it is that in far more than half of all retrodeviations the fixation
of the cervix anteriorly is present, always perceivable by the
sense of touch, and especially in case a Sim's speculum in the
knee-chest position is used, plainly noticeable to the eye. Contracting scars in the anterior vaginal wall, which in rare cases
may also result from vaginal lacerations in spontaneous births,
but for the most part are the consequence of premature forceps
deliveries undertaken while the os uteri was incompletely dilated,
operate, as a matter of course, exactly in the same way.   Only
if these scars are still fresh, at most three to four months old,
can they be stretched to such an extent that retrodeviation will
be cured.   If they are of longer standing, they are mostly too
firm and have to be excised.   But even in the latter case massage and stretching must be resorted to in order to prevent the
new scar caused by operation from contracting and again leading
to retrodeviation.

The only difference between my former view, respectively
that of Schultze, and the one which my investigations force upon
me, is, that the etiology which Schultze thought to be the most
common rarely or never occurs; that, on the other hand, the
one described on pages 127 and 128 of his book on displacements of the uterus as being produced by anterior fixation of the
cervix, represents the most frequent form of retroflexion.   In a
case illustrated in that connection (Fig. 55) likewise, the fixating
band is shown as running towards the foramen obturatorium
sinistrum, thus indicating that this is the most common form of
anterior fixation, although not especially pointed out in the text
of the book.

E. Martin propounds two theories: (1) Imperfect involution
of that portion in the uterine wall to which the placenta had

been attached and preponderance of the affected wall, so, that
in case the placenta had been attached to the anterior wall, a
retroflexion, or, if it had its seat in the posterior wall, an anteflexion would ensue; (2) the backward sinking of the puerperal
uterus by reason of its own weight during continued recumbent
posture. The first of these theories has been refuted anatomically
by Virchow, and, clinically, by Schultze, Von Winckel and many
other writers, who maintained that the causes of displacements
were to be looked for outside of the uterus, not in it; moreover,
it (the theory) did not account for the displacements in nulliparæ.
I should have preferred passing by this theory, because we can
not render any service to the name of a deserving gynecologist
by combating times and again an otherwise excusable error of
his, if not just recently made therapeutic propositions were
based on the opinion that the cause of the increase in the size
of the organ was to be looked for in the organ itself (orthopedy
of the uterus by tents, lessening of the angle of flexion in the
cervical canal by galvano-cauterization from the posterior vaginal
vault, manual bending of the uterus in retroflexion over its anterior
plane, etc.). Besides the hitherto adduced anatomical argu-
ments of Virchow, Von Winckel and others, to the effect that
the difference in the uterine wall was not the cause, but the consequence, of the displacement, and the additional causes of retroflexion, as clinically detected by Schultze, I have quite recently
observed cases that will furnish further material for combating
that theory. The cases in question were three of fixed retroflexion, in which the uterus had been successfully loosened from
the posterior pelvic wall under anæsthesia; but the uteri remained,
although forced by pessaries into anteverted position, bent over
their posterior side. At first this phenomenon might appear as
favoring Martin's theory; however, the layers of peritoneum that,
in this case, had produced this change in position, or, rather,

change in structure, in the uterus could plainly be seen. In
two cases I succeeded in correcting this defect by methodical
bending of the uterus over its anterior plane; in the third case
this anteversio cum retroflexione (as a counterpart to retroversio
cum anteflexione) continues, in case the uterus should lie for once
correctly in the pessary. Exactly thus it ought to be with every
uterus if, according to Martin's view, the cause of displacement
lay in its wall.

The second theory advanced by E. Martin, I am pleased to
state, I can corroborate with experience of the very latest date
with the same degree of justice with which I refuted the first
one. In two cases parametritis and endometritis were treated
by me before I remedied the partial perineal defect by means
of a plastic operation. After the patient had been in a recumbent position for scarcely three weeks, I found the uterus (*non-
puerperal*) displaced backwards. A few massage sittings were
sufficient to cure the displacement, but it would have been a less
easy task, perhaps even an impossibility, if a greater space of
time had elapsed since its appearance. For the future I have
made it a rule to limit, wherever not urgently required, the
recumbent posture of the patient in favor of sidewise position,
in which, according to Kuestner, the uterus gravitates towards
the front; and in cases in which the recumbent position was
especially indicated, to examine at the conclusion of the treatment for backward displacements.

We shall but briefly discuss the three additional causes
mentioned by Fritsch while treating of retroversion and retroflexion in Billroth-Luecke's manual, namely: (1) Fall upon
the buttock and consequent acute retroflexion; (2) overdistension of the bladder during childbed; (3) relaxation of all the
pelvic organs through masturbation.

Sometimes the symptoms of retrodeviation may be due to

a fall from a horse, violent coughing, sneezing, desperate efforts
to catch a firm hold of something while falling, severe straining
of abdominal pressure in lifting heavy burdens—all of which are
apt to produce a sudden backward displacement. But in such
cases, as Schultze says, we do not know positively whether these
displacements existed before the accident, without producing
any complaints whatever. The same holds good in the cases
observed by Fritsch. According to experiments made by Kuestner,* a large part of which I witnessed myself, the cervix
descended downwards and forwards through increased abdominal pressure, as showed by a little rod hooked into the cervix
and which emerged from the vaginal orifice to the extent of
2–3 centimeters. Theoretically it might be conceivable that
through such an approach of the vaginal portion to the anterior
pelvic wall, where, moreover, the abdominal pressure is directed
against the anterior wall of the uterus, and consequently the
corpus enters the Douglas space, a retrodeviation should ensue.
However, the aggregate results gathered by Kuestner from those
experiments are against such an assumption. Kuestner exposed
a number of puerperæ to such injurious conditions as were
hitherto believed to produce retrodeviations, as straining of
abdominal pressure, overdistension of the bladder and rectum,
lifting, carrying, etc. The result was that, even by combining
several such pathogenic conditions, retrodeviations occurred
only where such had already previously existed, or where, at
least, the parametria were in a diseased condition. Another
argument against the above theory is furnished by the result of
an experiment made by me. If the anteflexed non-puerperal
uterus is placed in retrodeviation, it will always resume spontaneously its anteflexed position. True, these pathogenic manipulations were not by any means as powerful as the harmful con-

*Zeitschrift fuer Geburtshuelfe und Gynækologie XI, 2.

ditions mentioned above. But how is it that no acute retroflexions ensue after operations performed on the vaginal portion
of the cervix, during which the latter is always drawn forwards
and downwards for a long time? Most likely such an acute displacement would heal, even if the normal position were enforced
for but a short time.

That overdistension of the bladder may cause retroflexion,
I can prove by a case from my own experience.

Mrs. W., in M., consulted me on account of a not entirely
aseptical perineo-vaginal laceration. On the fifth day of her
confinement I found, with some difficulty, the fundus uteri at
the lower margin of the liver, that is, it had been pushed up
in such a manner as an overdistended bladder will do, according to my clinical experience. Overdistension of the bladder
did not exist at the time, as was shown by the catheter, but had
in reality existed according to the statement of the intelligent
woman, although denied by the midwife. When after the healing of the wound, I made a bimanual exploration at the introitus,
the fundus was found lying retroverted and fixed to the right
posterior wall of the pelvis.

I could add many other cases, if necessary. In most cases
it is recumbent position combined with retention of urine, as in
childbed, after plastic perineal operations, etc. Under such
conditions displacements are apt to occur with a greater degree
of probability. The already heavy uterus is still more burdened
by the weight of a filled bladder and sinks backwards, mostly
to the right side; the bladder undermines, so to speak, the peritoneum, which up to the umbilicus is only loosely fastened by
scanty long fibers of connective tissue. If the bladder is not
soon emptied—ischuria paradoxa may often exist for days
after entirely normal birth—imperfect involution of the peritoneum of the anterior pelvic wall, not infrequently also of the

abdominal walls, will be the consequence, whence retrodeviation, often also pendulous abdomen.

In regard to relaxation of the supports of the uterus by onany, I have become in general distrustful of the doctrine of relaxation, consequently also of that caused by masturbation. In prostitutes, who are the victims of a similar abuse of the genitals, nothing of the kind is found. I observed such labile uteri repeatedly, but in such cases there was always inflammation in the parametrium, which fixed the uterus. If the bladder was not overfilled and if the patient voided it in an inclined position, the uterus was subsequently found anteflexed; but if, by an overfilled bladder, the uterus had been forced backward and downward, and if during the urinal discharge the conditions of gravitation were less favorable for the forward sinking of the fundus, the uterus would remain retrodeviated. In these cases to make a new test, I removed the fixating band and advised the patient to void the bladder diligently. A cure never failed.

Kuestner (l. c.) advanced two causes not found in other writers, namely, (1) laterally extending parametritis and (2) incomplete descent of the ovaries. He had observed the genesis of retroflexion—first, extensive parametric exudation forcing the uterus to the right, then resorption, thereupon shrinking, and finally, after months, retroflexion, whose mechanism of formation Kuestner could observe anew with each bimanual replacement. The parametric band had placed the uterus in an oblique position, so that the axis of rotation ran from the lower left to the upper right; the sinking backwards of the left cornu was thereby greatly facilitated, and became quite regular whenever the bladder was distended.

This modus, too, I tested experimentally and established fixation by means of needle and thread. The uterus can be placed in retrodeviation by filling of the bladder only when the

fixation of the cervix is at the same time directed downward.
The only effect of a lateral fixation directed more or less upwards
is that the corpus turns towards the opposite side. In my
pathologic-anatomical investigations I found formerly more frequently lateral parametritis as a cause than subsequently when I
for years carried on a clinical and critical review of these ques-
tions. For while carefully examining the fixations one may often
find at the same time fixations extending anteriorly towards the
obturator foramen (Fig. 12).

The mechanism of the formation of retrodeviation in such
cases is the same as when the cervix is fixed in front. With
every distension of the bladder the uterus is lifted so far as the
ductility of the tissues surrounding the cervix will permit; should,
however, the tissues be contracted or shortened laterally and
downward, and still more anteriorly and downward, the uterus
is, as it were, overrun by the bladder and displaced towards
the back.

At the beginning the peritoneum of the anterior pelvic wall,
under favorable conditions of gravitation, draws the corpus
uteri forward; later on it becomes loosened, flaccid, and as a
consequence powerless to do so any longer.

The incomplete descent of the ovaries, which Kuestner (l.
c., p. 53) regards as the probable cause of most cases of retroflexion of the virginal uterus, has likewise become since a subject of my pathologic-anatomical as well as clinical studies.

The ovaries, as is well known, are developed from the
genital ridge of the Wolffian bodies, and, like the testes in the
male sex, situated in the lumbar region, near the kidneys.
There they descend, like the testes, downward, or, if Cleland's
theory concerning the descent of the testicles is correct, the
other body grows upwards from the gubernaculum Hunteri. In
the female the structure corresponding to the gubernaculum

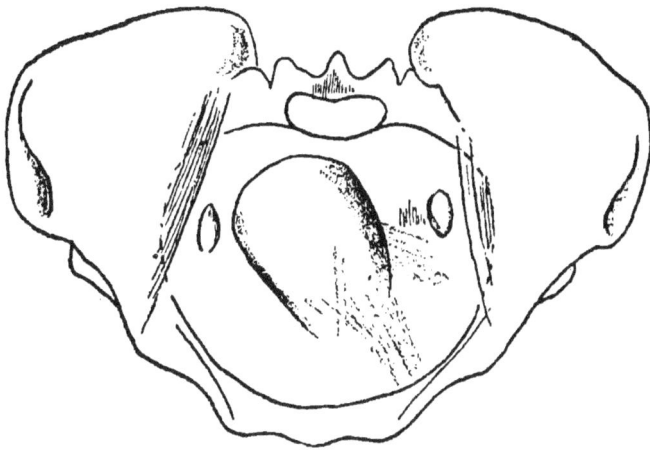

Fig. 12. Retroflexio uteri produced by anterior and lateral fixation of cervix
⅓ nat. size.

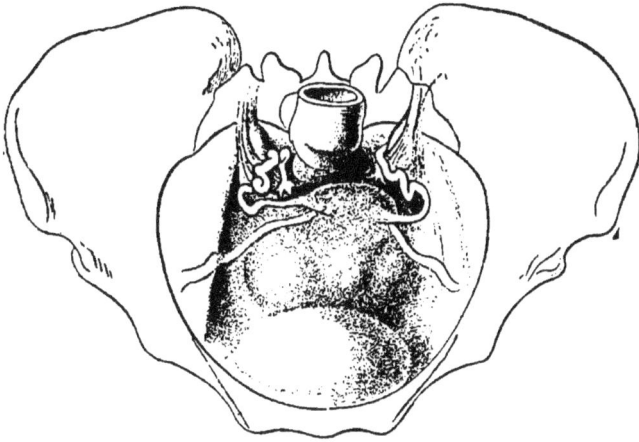

Fig 13. Retroversio uteri through incomplete descent of the ovaries, etc.

testis and from which subsequently the ligamentum rotundum and ligamentum utero-ovaricum are developed, must needs have been much longer originally, and, contrary to the process in the male, has grown somewhat simultaneously, for the whole process here is effected much more incompletely, the ovaries having advanced only as far as the pelvis. If this process is checked, the testes not being allowed to descend into the scrotum, we have cryptorchism in the male; while in the female we find retrodeviation of the uterus, the ovary drawing the fundus backwards by the ligamentum ovarii.

My pathologic-anatomical researches have confirmed the correctness of these statements, having described (l. c. p. 51 and 52) three such cases. Two of these, a girl of three years and another of sixteen years, showed the above condition only on the left side, while in a girl fourteen years old both sides were found to be thus affected (Fig. 13). The ovary which had failed to descend had an elongated, spindle-like form. In the case of the sixteen-year-old girl its upper end lay 3 cm. above the greatest prominence of the promontory, in the one fourteen years of age both ovaries were at the same height, the distance of the two organs situated at the side of the promontory being only 4½ cm., and both were bent over the linea terminalis. In the child I was looking for causes, and I thought I could perceive traces of fetal peritonitis above the ovary. The course of the ligamentum utero-ovaricum was in all cases straight and stretched and therefore prevented the replacement of the uterus,

-a proof that it was the cause of the deviation. The form of the approaches to the space of Douglas was in all cases correspondingly changed. In the last mentioned case it was a small rectangle. It was the most pronounced case of the kind; that of the child the least conspicuous, although the uterus was distinctly

retroverted. In his " Textbook of Women's Diseases," p. 337.
Von Winckel relates four cases of his own and two as described
by Ruge of retroflexion in the child. Unfortunately nothing is
said about the ovaries.

Since that time I have examined my patients with this
special question in view, and I cannot admit that (incomplete
descent of the ovaries) is the most frequent cause of retrodeviations of the virginal uterus. Among the numerous retrodeviations
observed by me there were only six cases in which a condition
could be found analogous to the one described. In three cases
the arrest of the development was doubled-sided, in the other
three only one-sided. In the first three cases, in spite of great
lateral mobility, bimanual replacement was not possible, because
on account of the narrow approaches to the space of Douglas
one could not get behind the uterus. Two of these were women,
consulting me for sterility, both otherwise free from any ailments; one was a virgin and came to me simply on account of
dysmenorrhœa.

Where the arrest of the development was only one-sided,
the uterus could be replaced. In two cases, by means of massage and stretching, the ovary could be drawn into the pelvis and
a cure effected, in the third case no such attempt was made. In
one case, however, there occurred a relapse, although the ovary
lay in the pelvis.

In numerous, less pronounced cases of incomplete descent
of the ovaries it is not a matter of impossibility that, during the
whole period of childhood, the uterus keeps its normal position
in the pelvic axis. Not until the age of puberty has begun and
with it anterior curvature in the lumbar region, whereby the
whole pelvis is, as it were, anteflexed and thus assumes a
stronger incline, is the utero-ovarian ligament strained and the
fundus drawn over beyond the axis of the pelvis. The great

majority of retrodeviations of virginal uteri, however, is not congenital, but acquired and brought about by one or the other of
the causes mentioned.   That, in principle, it does not matter
whether only the uterus is virginal or the entrance of the vagina
also, goes without saying.   In all cases of incomplete descent of
the ovaries anatomically examined by me, the anterior vaginal
wall was found to be very short, so that it seemed as if the
etiology given by Schultze and Kuestner covered the same cases.
Still I have since examined cases of retroflexion and short
anterior vaginal wall in which the ovaries were situated in the
pelvis.

The cases in which parametritis superior or spermatica, after
drawing the fundus to the side, continues to contract so as to
draw the fundus uteri backwards and upwards, closely resemble
those just described (see Fig. 10).   While there the descent of
the ovary was incomplete, here it is again drawn upwards by the
contracting perivascular connective tissue band, and the same
may happen to the uterine end of the tube, whence the ovary
starts; the fundus is then situated considerably higher up than
usually, and on a hasty examination one is inclined to believe
that he is confronted with peritonitic fixations.   Often the ovary
also is inflamed, since the ovarian vessels originate in the spermatic vessels, branches of the spermatic or ovarian arteries
being sent out to the uterus and tubes and anastomosing also
with the uterine arteries.   Ovaritis may thus lead to real peritonitic adhesions between ovary and tube, etc., and render the
deception more plausible.   Do these cases coincide with those
described by Schultze as high backward fixation of the cervix
with elevation of the fundus?   Both illustrations and remarks
about their frequency seem to justify such an opinion.   Schultze
observed the displacement eight times in 134 cases.   But the
cervix is only frequently fixed, not always.

Von Winckel's view, according to which chronic obstruction due to continued distension of the rectum, in connection with intense abdominal pressure exerted during evacuation, induces retrodeviation, makes the impression at first as if it were embodied in previous opinions. At one time it seems to be contained in the theory of Fritsch, at another time in the doctrine of the relaxation of the peritoneum of the posterior pelvic wall, and yet the modus of the formation of retrodeviation is quite an original one. By undue exertion of abdominal pressure, according to Kuestner's experiments, the cervix is forced forward, as also by the succeeding passage of the fecal mass which often stagnates in the rectum for a long time. Gradually the tissue in front of the cervix becomes condensed and retrodeviation stable. Obstruction being seldom the cause, much more frequently the consequence, of inflammations in the pelvis, this mode of originating must be rare. I cannot speak from my own observation.

I shall proceed now to discuss Thure Brandt's theory, upon which the essential part of his therapeutics, the liftings, is based, namely, relaxation of the round ligaments. In his publications this theory can be read only between the lines, but by way of epistolary correspondence it has been fully and clearly explained to me. The Alexander-Adam's operation, respectively Alguie's operation, the shortening of the round ligaments, ought to be based on the same theory, if it rests at all on etiological considerations.

As early as 1887 this view has been refuted by the pathologic-anatomical investigations and measurements made by me. Let us quote that passage, l. c. p. 34:

" In order to gain some positive knowledge about the functions of these so-called bands, I tried to determine whether retroflection of the uterus affected the ligament qualitatively or quantitatively.

" In nearly all cases that came under my observation I first measured the distance of the fundus uteri from the symphysis and thereupon the length of the ligament and its thickness about in its middle.   In thirteen anteflexions I found the distance between the fundus uteri and the posterior margin of the symphysis to be 7 cm. on an average, the maximum 10 cm., the minimum 5 cm.   In nineteen retroflections the mean distance was 12.43 cm., the maximum 15 cm., the minimum 9.5 cm.

" The length of the round ligaments in fourteen anteflexions (twenty-eight measurements) was 18.5 cm., the thickness in twenty-four measurements (twelve cases) 4 mm.

" With these data I compared the corresponding measures in retroflexions—18.77 cm. in length and 3.77 mm. in thickness.

" The result was a difference of 0.27 cm. in length in favor of retroflexion and one of 0.23 mm. against retroflexion.   One might be tempted to assume that this result was due to an ever so small stretching of the ligaments.   But among the cases of anteflexion there was one of metritic enlargement (see Fig. 10, p. 72), in which the length of the round ligaments amounted to only 15 cm., but the thickness to 7 mm.   This case sufficed to affect the calculation in the sense indicated.*

" In the nineteen cases of retroflexion the length of the round ligaments was sixteen times unequal on both sides, equal only four times. In eleven cases the thickness, too, was unequal, and in nine of these eleven cases the longer ligament was at the same time also the thicker one.   Only in two cases the shorter ligament was thicker.   This certainly does not favor the idea that the ligaments become thinner by stretching, but bespeaks rather a possible hypertrophy of the stretched cord.   However, the figures from which the averages are calculated are too small

* If we leave out this case, the length of anteflexion is 18.54 cm., the thickness only 3.73 cm.

to preclude an accidental coincidence, especially as an analogous
relation in cases of anteflexion could not be established."

On page 35 we read:

" For the rest, the ligament sympathizes with the uterus,
as if it were part of its wall.   In case the uterus is hypertrophied,
hyperæmic, thick, the ligament is similarly affected; if the uterus
shows varicose veins, the ligament likewise discloses dilated vessels.   Most likely, too, the ligament contracts only when the
uterus contracts, that is, during puerperal labor.   A similar
responsive sympathy may be supposed to dwell in the prolonga-
tion of the uterine muscular tissue towards the back.   Only
when both are pulled do they exhibit labor hypertrophy (in
prolapse)."

In this connection it must be especially emphasized that
origin and insertion, on an average, are only 7 cm. apart, while
the length of the so-called bands is 18.54 cm., or well nigh three
times as much, their course being very tortuous.

They are consequently not even strained in retroflexion,
although the distance between origin and insertion amounts to
12.40 cm. on an average.   Accordingly, length (18.77 cm.) and
thickness (3.77 in the mean) do not differ to any extent worth
mentioning.   If we expect any effect at all from the operation,
the so-called ligaments must be shortened 12 cm.   In this case
broad peritoneal folds would be drawn from the right and left
over the pelvic entrance.   Gradually they would be lengthened
again, the uterus extricating itself from the muscular structure.
The round ligament receives its muscular fibers from the muscular coat of the uterus; in stretching them the muscular bundles
are drawn out in a fan-shaped manner.   The lowermost bundles
almost reach the external os.   The muscular bundles of one
side form then a rhombus with those of the other side, described
already by Rokitansky, sen.   By hygienic gymnastics these

round ligaments are therefore in no way to be shortened, or, in other words, to be stretched so as to gain an effect upon the position of the uterus.

Through his liftings Brandt accomplishes something entirely different from what he intends to achieve, namely, the stretching of cicatricially contracted connective tissue, which fixes the cervix pathologically in front and downward. The round ligaments are nothing else but the remnant of a fetal structure, corresponding to the gubernaculum Hunteri in the male, and which, as serous cysts occasionally encountered in it indicate, seems to have been originally a hollow structure. The uterus sends out muscular prolongations in all directions. A very strong, long bundle, much stronger than that of the fold of Douglas, for instance, follows the spermatic vessels; shorter ones run from the cervix to the cul-de-sac and along the entire lateral border to the peritoneum of the fossa Claudii. All these muscular prolongations are dependent, contracting only when the whole uterus contracts itself (during labor pains, colic after intrauterine treatment). One-sided contraction, for instance, of one of the Douglas folds or of a round ligament, belongs into the realm of fiction, for the reason alone that every one of these so-called bands would need a set of nerves of its own.

Now there are cases in which the cervix is situated plainly in front of the spinal line and therefore fixed in front, and yet the uterus is not in retroflexed position, but in anteflexion. Such cases conflict with the above given theory only apparently. Evidently cystitis was the cause of the anterior fixation. This inflammation induced also inflammation and shrinkage of the peritoneum over the bladder, thereby fixing the cervix and corpus anteriorly. The Alexander-Adam-Alquie operation has as its effect just the same position of the uterus, not the normal position. In both cases bladder troubles are always present.

10

Retroflexio uteri fixata may take its origin from any of the
above mentioned causes. Where the fundus uteri rubs against
the posterior pelvic wall, both peritoneal surfaces become chafed
and adhesions form. Of course, we have to assume that the
uterus is moved by exertion of abdominal pressure, bowel movement, etc., in order to produce a certain exasperation and that,
on the other side, it be not moved too much nor too long,
because otherwise the plastic exudate will be ground down
again and adhesions will not form. In the last two years four
fixed retroflexions formed, as it were, under my eyes. The last
case was the one of the puerpera mentioned above (page 160), in
which overdistension of the bladder was the cause of retrodeviation. The temperature had been taken very carefully, never
had there been any fever, and yet the uterus was found to be
adherent posteriorly at the right side near the promontorium.
It seems that in this case the labor pains favored the exasperation and the succeeding rest the agglutination. In another case
inflammation had most likely existed previously, subsequently
inducing exudation and agglutination. Another puerpera had
post partum hemorrhage and high fever on the tenth day of her
confinement. Two colleagues of mine having curetted the
uterus without any success, called me into consultation. I
made a manual exploration and removed, by means of a placental forceps, a sapræmic placenta succenturiata (the first
placenta had been examined by the physician and found to be
complete). The bleeding stopped at once and the fever sub-
sided gradually under antiseptic irrigations. Three months later
I found a fixed retroflexion of the uterus. That the uterus,
owing to its own weight and to continued recumbent position,
had sunk backward, could be inferred from the fact that,
slightly concave anteriorly, it had adapted itself, as it were, to
the posterior wall of the pelvis.

Two other cases had been caused by retrouterine hæmatocele. The blod clot in the cul-de-sac had forced the cervix forward and upward and the fundus was in retroflexed position.
In both cases—and that is most essential—I was well acquainted
with the previous condition and knew that the uterus had been
anteflexed. Cases of hæmatocele complicated by retroflexion,
on account of their frequency, have in all probability been met
with by every gynecologist. The effusion having been absorbed,
the uterus remained in both cases in fixed retroflexion. In
order to ascertain why the uterus in many cases of retrouterine
hæmatocele is only in anteposition, in others in retroflexion, I
measured in fifty-seven cases the depth of the cul-de-sac, that
part of the Douglas space which, above, is bounded by the
Douglas folds and is usually not unfolded, forming only a capillary fissure, but in which an intraperitoneal effusion collects at
first. According to these measurements the depth of this space
was found to fluctuate between 1.5 and 12 cm, so that consequently an effusion may extend just as far down into the
recto-vaginal septum. As the latter underneath the vaginal portion of the cervix bends forward and up to its exit through the
levator ani muscle runs horizontally in the pelvis, the effusion
in case of a flat cul-de-sac is found only behind the cervix and
causes only anteposition, but in case of a deep cul-de-sac also
underneath the cervix. Under such circumstances anteposition
and elevation of the cervix bring about retroflexion, and the
peritonitic adhesions remaining after absorption of the exudation cause the fixation.

In conclusion, let me mention a form of genesis of retroflexion which hitherto I knew only from pathologic-anatomical
investigations. but which nevertheless appears to be quite common, being found frequently in a proportionately small number
of anatomical specimens. Clinically, it can hardly be diagnos-

ticated, but it may be assumed whenever it is impossible to find
another cause for retrodeviation.

It occurs when the peritoneum of the vesico-uterine excavation becomes adherent to the uterus below its normal attachment.   This normal attachment of the peritoneum to the uterus
begins *not lower than at least 1 cm. above the internal os.*   In
case the bladder becomes distended, the peritoneum is lifted up
to that point from the uterus, because the bladder, having
become larger than normal, needs more peritoneum for its cover.
The corpus uteri is thereby drawn forwards and upwards.   If
the bladder is entirely empty and small, the distance between
its anterior border and the cervix is often 2.5 cm.   The surplus
portion of the peritoneum lies in the excavation in loose proximity to the anterior vaginal vault and the anterior wall of the
cervix.   Inflammation in this region, the bladder being continually empty on account of violent vesical tenesmus, leads to
adhesions of the peritoneum to the cervix.   As soon as the
badder becomes more distended, the peritoneum, instead of
pulling on the corpus uteri, pulls on the cervix uteri, and retroflexion ensues.

Viewing all these complicated modes of formation of retroflexion from the standpoint of broad, general principles, the
etiology becomes simple enough: Displacement of the cervix
forward and downward, in rare cases also forward and upward,
and displacement of the corpus uteri backward, cause retrodeviation, however different the displacing forces may be.   The
duty of therapeutics therefore consists simply in removing those
alterations in the pelvis which displace the cervix forward and
the corpus backward.

TREATMENT.

After an exhaustive exposition of the etiology, which may
have appeared too minute to many a reader, I shall be more

brief in setting forth the treatment. The latter can be condensed into a few words: *Massage* and *stretching* of those fixations that hold the uterus in an abnormal manner. The principle remains always the same, in whatsoever manner we may
direct the force upon the abnormally fixed organ. A vast number of various methods might be distinguished, if we wanted to
describe as a separate method each particular way of applying
the fingers or stretching with the aid of any instrument. The
uterus having always to be used as a handle, if we wish at all
to stretch abnormal fixations effectually, bimanual replacement
of the uterus is the beginning of every treatment calculated to
cure a displacement. It is true, in rare cases, especially of
retroflexion, massage must be resorted to on account of the
sensitiveness of the peritoneum until the sensitiveness is
diminished and the abdominal walls have become relaxed.
Thus every effectual treatment begins with replacement of the
displaced organ.

Instrumental replacement is not advisable, because it produces pains and wounds in the uterus that might be easily
infected during the treatment. *Marion Sims* already gave utterance to this sentiment, in spite of construction of his repository,
and advocated a method of three sponge-holders as the better
one. Although it was less perfect, it nevertheless perfectly coin-
cided in principle with the method of Schultze.

Schultze and Brandt are surprisingly unanimous in reject-
ing any instrument whatever for the purpose of replacing a displaced uterus. Schultze allows only one exception, claiming
that replacement by sound is permissible in case one has convinced himself, eventually by the aid of an anæsthetic, that the
uterus can be replaced and that only an excess of adipose tissue
in the abdominal walls or an exceedingly large pelvis render
replacement difficult, it being impossible to make use of anæsthe-

sia every time one wishes to treat the patient.   Only twice I
was obliged to resort to the sound in replacing the uterus.   In
one of these cases both assumptions of Schultze concurred, in
the other case I had to deal with a highly nervous and indocile
patient, unable under any circumstances to relax her abdominal walls.

METHODS OF REPLACING THE UTERUS.

Brandt sets up a rather complicated system, according to position and the points of attack in which, or rather from which, the replacements are attempted.

A. Replacement in upright position, or rather abdominal posture:

    1. Recto-vaginal replacement.

B. Replacement in crook-half-lying position:

    2. Ventro-vaginal replacement:

        (*a*) Tipping or turning over (Umkippung).

        (*b*) Hemming or squeezing in (Klemmung).

        (*c*) Hooking-on (Einhakung).

        (*d*) Replacement pressure (Repositionsdruck).

    3. Ventro-rectal replacement.

    4. Ventro-recto-vaginal replacement.

Appendix: Replacement with one finger.

This complicated system is very much simplified by conceiving the position as unessential to the method and only incidentally mentioned, and by distinguishing only a different adjustment of the hands as a particular method.

Thus 1, 3 and 4 may appropriately be compressed into one in the description.

The index finger of the left hand is inserted into the rectum as high up as possible, at least up to and above the ampulla and beyond the narrow space where the folds of Douglas are situated and the so-called sphincter tertius is supposed to be. The thumb is introduced into the vagina, and, after the fundus has been gradually and cautiously lifted out of the hollow of the sacrum and pushed forward with the index finger, the uterus is grasped with both fingers and turned towards the front. The latter manipulation is best effected by lowering the elbow as much as possible and by means of a vigorous bending of the wrist. The right hand resting upon the abdomen now endeavors, beginning at the upper part of the abdomen, to distribute the intestinal coils by gentle circular rubbings, and gradually to penetrate deeper and deeper behind the fundus and to push it entirely over to the front.

In cases in which the uterus is very large and heavy, as in puerperæ and pregnant women, abdominal posture, or still better, knee-chest position is to be preferred, because thus the heavy uterus is made to gravitate forwards and can be pushed more easily out of the hollow of the sacrum and forwards. In those cases in which our fingers prove to be too short or the uterus is situated too high up, the first part of the treatment— the lifting out of the hollow of the sacrum and the forward directing of the uterus—may be performed in an upright position, because in this position the uterus is forced down by abdominal pressure. The concluding portion of the manipulation, complete replacement anteriorly, is done always in semi-recumbent position (crook-half-lying). In virgins or in women with very sensitive vaginal entrance, the thumb should not, if possible, be inserted into the vagina, and only that method of replacing should be employed which, lower down, is described as hooking-on (Einhakung).

If we use anæsthetics and introduce two fingers into the rectum, and if, besides, we apply a vulsellum or a Muzeux forceps to the vaginal portion in order to draw the uterus down, we employ a method which Schultze recommends as the best one for diagnosing the seat of large abdominal tumors. If we use

a greater amount of energy in pushing the fundus uteri towards
the front, as much as it is at all possible under the circumstances, we employ the method advised by Schultze for loosening fixed retroflexions.

Accordingly, only *a*, *b*, *c*, *d* and appendix are left for discussion.

Tipping over, or tilting over (Umkippung), can be successfully done only in cases of rigid and straight uterus and elongated
vaginal portion of the cervix. It is executed by placing the
index finger, or, as I do, the index and middle fingers, with the
touch surface turned backwards, against the vaginal portion of
the cervix in the anterior vaginal vault and pressing obliquely
backwards and downwards. The fundus is thereby lifted forwards, the cervix being used as a lever. That this method is
attended with success only in those cases susceptible of easy
replacement, in which sometimes retroversion is not detected until
the second or third examination, is self-evident.

Hemming, or squeezing in (Einklemmung), will lead to a
satisfactory result, even in cases of more flexible uteri, provided
only the vaginal portion is long and the cervix somewhat movable, so that by pressure exerted upon the angle of retroflexion
backwards and downwards the uterus can be pressed flat against
the sacrum and there firmly held. If, now, under light circular
rubbings, the right hand is placed on the abdomen above the
fundus in such a manner that the ball of the hand rests upon or
beside the promontorium, and that the finger-tips, pointed downward towards the pelvic floor, rest close to the fundus; and if,
with the fingers of the left hand, the uterus is pushed along the
sacrum upwards, we often succeed in forcing the fundus upwards
in front of the right hand and replacing it, bimanually, in its
anterior position.

Hooking-on (Einhakung) is attended with success although

the above named favorable conditions may be entirely absent, provided only that the sensitiveness is not great and the abdominal walls are pretty well relaxed. The fundus, which in case of flexible uterus lies deeply imbedded in the excavation, is, with the fingers of the left hand, at once lifted from the posterior vaginal vault so high above the pelvic entrance that, proceeding from above, under gentle circular rubbings and penetrating into the depths, one can get with his right hand behind the right tubal cornu of the uterus. It is then a comparatively easy matter to hook on with one's fingers, to lay the cornu and finally the whole corpus uteri over to the front and to force, at the same time, with the inside fingers, the vaginal portion of the cervix towards the back.

This manner of replacing the uterus coincides exactly with that given by Schultze. I employed it for years to the exclusion of all others before I became acquainted with Brandt's methods and I always accomplished satisfactory results. True, I often was obliged to make use of anæsthetics. The same treatment, aided by anæsthesia, is distinguished by Schultze as a second especial method. Short fingers or a long vagina oblige the operator to insert a vulsellum into the vaginal portion, to draw the uterus down, and thus more easily elevate the fundus—a happy modification added by Kuestner to modus operandi.

Replacement pressure (Repositionsdruck) finally may lead to a successful end in cases where all the other hitherto mentioned treatments utterly fail, even if all unfavorable conditions, such as short anteriorly fixed cervix, narrow vaginal vault, flexible and soft uterus prevail, so that, as Brandt expresses himself, the fundus seems to be glued to the posterior pelvic wall. This method is executed in four manipulations: (1) The intravaginal fingers lift the corpus in the posterior vaginal vault as high up as possible and hold it there. (2) The right hand, under circular rubbings, is steadily penetrating deeper into the pelvis, until it is able to level out the angle of flexion and, with the finger tips, to press that portion of the uterus against the posterior pelvic wall. (3) The fingers in the vagina are placed in the anterior vaginal vault over the same spot (the point of flexion). (4) The finger tips of both hands push the uterus along the sacrum upwards to the pelvic entrance. Here the right hand, proceeding from above under circular rubbings, endeavors to get behind the right tubal cornu, or rather the fundus, and lay it completely over to the front.

Brandt's advice, to begin high up on the abdomen and to advance gradually towards the fundus, is very much to be heeded. The abdominal walls, as far as they are movable, must be taken along up to the spot where we wish to place the hand in beginning the circular rubbing movements, so that we need not displace the hand upon the abdominal wall. If we advance straight from the front towards the fundus, a backward pressure is imparted to the latter through the medium of the intestinal coils, enabling it to escape our fingers in a majority of cases. If, however, we advance with the hand from a point higher up, this pressure on the part of the intestinal coils and abdominal walls ensues likewise from above and imparts the uterus an impulse to move forward.

Not so much as an effectual method, but rather as a daring sample of his technique, Brandt adds to his description "replacement with one finger." In principle, it imitates replacement pressure, consisting in: (1) Elevation of the fundus; (2) turning over the hand toward the left side and simultaneous pressure upon the vaginal portion of the cervix; (3) under increased pressure of the finger from the right side against the vaginal portion, withdrawal of the finger and placing of the same in the anterior vaginal vault; (4) upward forcing of the cervix to such a height that it is pushed up behind the corpus, compelling the latter to escape anteriorly. That a long finger is needed for this manipulation, such as is allotted to but few mortals outside of Brandt, can easily be perceived.

As a matter of completeness, we add the description of a method of replacement mentioned by Schultze and used by him in cases in which, on account of hemorrhages, he had dilated the uterus so far as to permit digital examination. During this examination he would bend the index finger, with the tactile surface turned forward, loosening with the hand any possible adhesions. I, myself, employed this method only once, Schultze only in a few cases. I particularly mention the latter fact because certain parties, not exactly kindly disposed, have, ex cathedra, designated this method as the one most frequently used by Schultze.

The uterus having been replaced, the next exploratory step consists in looking for adhesions that might draw the uterus back again. Sensitive portions have to be massaged, and *adherent ovaries to be loosened by massage and brought forward.* In case thick layers of peritoneum are present on the posterior wall, the uterus must be energetically bent over its anterior plane. Above all we have to search for parametritic fixations of the cervix, which are apt to fix the latter in front, at the side, and downwards, or for such as fix the corpus upwards and backwards. The fixating band having been accurately diagnosed, its sensitiveness is removed by means of bimanual massage. Stretching causes renewed pains and swelling, which in their turn have to be dissipated. In lateral fixation we apply the fingers of both hands on that side where the band is found, and

---

*In a great many cases retrodeviation of uterus is complicate 1 by descensus (prolapse) of one or both ovaries, lying either against the side of the supra-vaginal portion of the cervix and body of the uterus or at the bottom of the cul-de-sac of Douglas.—The Translator.

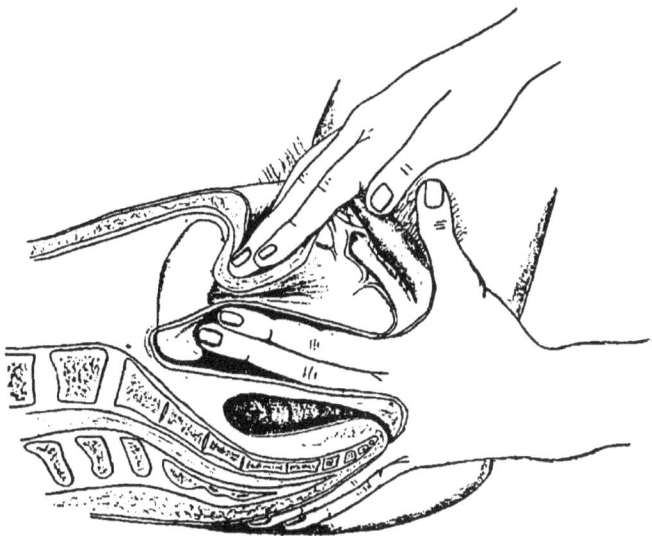

Fig. 14. Stretching of ananterior fixation of the cervix in case of retro flexion of the uterus.

push the uterus towards the right if the band is on the left side, or draw it to the left if the band is on the right side, at the same time endeavoring to elevate the uterus as much as possible.

If the fixating cord extends towards the front, we place the fingers of both hands in the anterior vaginal vault and push the uterus backwards and upwards (see Fig. 14); but the promontorium being in the way if we press directly backwards, and, moreover, the band running obliquely in most cases, it is better to stretch and elevate in the direction of the oblique diameter.   When the pelvic connective tissue is shortened more diffusely, we stretch sometimes in straight direction, at other times in the direction of this or that oblique diameter, the fundus being again and again pressed forward and downward.

Should superior parametritis be the cause of retrodeviation, the most important thing to do is to stretch the band, usually running from the tubal cornu to the right, backwards and upwards.   As the case may be, we draw the fundus to the left, forwards and downwards; likewise the whole uterus in case of total parametritis.

It is this very form of retrodeviation in which I never experienced any failure, provided only massage and stretching were continued for a sufficient length of time.

In two cases of incomplete descent of the ovaries I succeeded in bringing the ovary down into the pelvis and in effecting a cure, the arrest of the development having been one-sided. In the remaining cases I did not attempt a cure.   The manipulation is the same.

As a matter of course, the massage sittings must by all means take place daily and are not to be interrupted during menstruation.   Just at this time the cicatricial contractions are most ductile.   In this way I continued the treatment for weeks, formerly even for months, until every fixation had disappeared

and the uterus had regained its normal mobility. A great number of retrodeviations, about one-half, were cured.

In cases where there was no dysmenorrhœa, no discharge,
nor erosion, and accordingly no chronic endometritis, I concluded the treatment, just as Schultze is waiting to see, after
the uterus' lies correctly in the pessary, whether the fluor will
not disappear by itself after the disappearance of the torsion of
the broad ligaments. In case of chronic endometritis, however,
we treat the latter, and on this occasion push the cervix seized
with the vulsellum backwards and upwards. Finally, yet, the
usual two or three massage sittings, in order to remove any
possibly recurring pains. Thus all troubles will have disappeared, even in cases where the uterus does not lie anteflexed
after the treatment.

In this manner I had cured many a retrodeviation, the first
one in the year 1887, and I thought already that I had discovered a new method (I had seen Brandt's own manner of treating at Jena in the year 1886), when, fortunately enough, before
the intended publication, I visited Brandt, in February, 1888,
for the purpose of ascertaining, then and there, all about the indication, hygienic gymnastics, etc., what Brandt really intended
by every single part of his therupeutics. On this occasion I
found out that my method had already been discovered,—by
Major Thure Brandt himself.

Only when the uterus is replaceable, approximately normally movable, and only when there are no more surrounding
inflammations and bands that can be stretched easily bimanually,—only then, I repeat, must lifting movements be executed.
I am sorry for certain inventors that their methods had already
been anticipated, and a third one could have saved himself the
trouble of arrogating to himself the right of priority.

It is a rare and often admired gift of Brandt, a certain

unconscious feeling, which led this unusual man to the right
goal, without his being able to comprehend scientifically the full
import of his modus operandi.

Another question is whether these lifting movements are
necessary. Fortunately, I am able to say "No." I say "fortunately," for in the whole therapeutics of Brandt, there is
nothing so unæsthetic, nothing so laborious, nothing so hard to
learn, nothing so expensive, in short, nothing so disagreeable
for both physician and patient, as this lifting. Nothing was
more hindering to the dissemination of Brandt's method, nothing detained and deterred the physicians so much from its application. When, on Whitsuntide, 1889, before the congress of
gynecologists, I demonstrated the original method of Brandt in
treating prolapse by means of sacrum-beating (Kreuzbeinklop-
fung), alternate twisting (Wechseldrehen), knee parting (Knietheilung), pressure of plexus pudendus, in short with all its additional detail, there arose general merriment among the gynecologists present. Unfortunately, I did not have the opportunity
of speaking a second time nor of stating publicly that I, unlike
Brandt himself, was not exactly convinced as to the indispensa-
bleness of many a "mystic accessory detail," as Privy Councillor
Olshausen aptly remarked. But I believe that my paper in
"Volkmann's Vorträge," already in print at that time, shows how
earnestly and scientifically I endeavored to separate the essential
from the unessential. Incidentally stated, it would be asking too
much of gynecologists of the first rank engaged the whole day in
lecturing, operating, applying of pessaries, penciling of the cervix,
prescription-writing, giving balneological advice, etc.—who have
not time enough to attend to the crowd of calling patients and
receive such fees for their services as cannot possibly be
increased, to burden themselves with a treatment so laborious
and requiring so much time and at least some experience. Dur-

ing the time occupied by a single massage sitting they are able
to place five rings or pencil six cervices, and yet feel not so
fatigued as in executing lifting movements. Scarcely anybody
is able to pay a sum even only partly compensating a man so
much occupied for the loss of time involved in a sitting. How
much would the uncertain cure of a case of retroflexion cost?

It is left for younger physicians and gynecologists, who
have more time at their disposal, to take up this theory (of liftings). But these, for sundry reasons, are not able to keep any
female attendants, or hygienic gymnasts, to administer the
liftings or other necessary movements. In Munich, for some
time, a number of physicians united and assisted each other
mutually, but these, too, abandoned the idea as soon as the
ardor of their scientific experiment became dampened and the
time for the sober application of the practice was due.

Although, as stated above, I cherish the conviction that
lifting (as a hygienic movement) may be dispensed with, I nevertheless feel bound to describe it at length.

Lifting.

Lifting, Swedish lyftning, it was which mainly induced
Brandt to treat women's diseases.

A soldier while marching sustained a prolapse of the rectum.
At that time Brandt was an active officer in the army and assistant teacher of hygienic gymnastics at the Central Institute in
Stockholm, Sweden. He desired to do something for the afflicted
man, but did not know how to render assistance. Presently he
resolved to execute a resistance movement on the intestine supposed by him to have become relaxed—in other words, to stretch
vigorously the longitudinal muscular tissue in order to produce
a reactive contraction. He took his place at the right side of
the man, who was lying in a recumbent position with his thighs

flexed upon the abdomen, and with his left hand, supported himself lightly upon the man's right shoulder, while his right hand, palmar surface turned towards the patient's head, and under vibratory movements, beginning from the left iliac region, penetrated gradually deeper and deeper into the small pelvis. With curved fingers, under rather energetic counter pressure, he then pushed the bowel upwards. The man was instantly and permanently cured, and the S. Romanum lifting was discovered.

Since that time Brandt has used it in many cases of prolapsus ani and proctocolpocele. I myself have no experience in this direction, but as this book is intended not merely for gynecologists, but also for general practitioners who are likely to meet with such cases, I felt myself obliged to mention the method. I demonstrate it to my auditors every semester, and only a short time ago I became again convinced that the anus is in reality drawn in during the lifting.

This success achieved in treating prolapsus ani induced Brandt to try the same manipulation in the prolapse of the uterus, in order to bring about, by this kind of resistance movement, a contraction of the muscular tissue of the relaxed vagina and of the round ligaments of the uterus. Not until many years later could his fond idea be made to assume practical shape. The very first case proved to be a brilliant success. A prolapse which had persisted for more than a score of years remained in replaced position after the first lifting, although at that time the lifting was given in a rather primitive way, first with one hand, subsequently with two hands, and without any assistance. Probably the sensation of sinking, complained of by most of the patients, soon induced him to regard every case of retrodeviation as a new prolapse, in short, to make use of lifting wherever this feeling prevailed. This is the reason why Brandt is still using these liftings in many cases of anteflexed
11

uterus, also in pregnant women, and distinguishes this as a special kind of lifting, being of easier application. Pains and swellings he removed by massage according to the rules and principles usual already at that time with hygienic gymnasts. However, he soon found out that it was not so easy a matter to grasp the uterus always in every lifting, and he was obliged to control and support this bimanually.

The etiologically close relation he supposed to exist between retrodeviation and prolapse, a view shared also by Schultze (l. c.), induced him to treat retrodeviation and prolapse in exactly the same manner, likewise cystocolpocele with sinking. Not until later did he conclude to execute the lifting in retrodeviation in a way different from that used in prolapse. In both kinds of lifting he places the finger of the left inside hand, as well as those of the outside hand, in front against the movable uterus, to be replaced, the ball of the latter hand turned upwards, the finger-tips downwards, the touch surface backwards. In this way he shows the female attendant (gymnast) the spot where the vesica-uterine excavation is, at same time pressing the uterus with both hands as far back as possible. The attendant now places the feet of the patient together, with knees apart, and takes her position beside the patient's buttock in such a way that with one leg she stands upon the floor, with the other one she kneels beside the buttock of the patient lying in crook-half-lying position, so that the lower legs of the patient lie close to the attendant's inguinal region and the feet hang between her thighs.* The attendant, having placed the finger-tips of both her hands, with the ulnar margins together, the balls turned headwards and the touch surface backwards, upon the outside hand of Brandt, and bent herself forward, the latter withdraws

---

*The patient rests upon an apparatus (low cot) composed of two plane surfaces meeting each other at an obtuse angle.

his outside hands, and now the gymnast's fingers, straightened out, and with elbow joints stretched, endeavor to penetrate, parallel with the symphysis, in an oblique direction downward into the pelvis. In a number of cases the patients have to be asked to breathe deeply and to relax the abdomen. Presently the attendant rises, and while doing so presses vertically downwards into the pelvis. She then curves her finger-tips and elbows upwards, and, while rising further, pushes the uterus upwards along the sacrum, beyond the promontorium, in short, as far as feasible, until finally the uterus glides gradually away from the fingers. This is the proceeding in case of prolapse.

In retrodeviation the lifting is begun exactly in the same way, but as soon as the uterus has been forced so far backwards that part of the tissue extending from the uterus towards the front ("the anterior holding parts") is tightly stretched, Brandt calls a halt. For a few seconds longer the uterus is being kept in position until the hands are removed, pretty rapidly, in a horizontal direction, i. e., in case of a woman in lying position, vertically upwards. The latter movement imparts the fundus an additional impulse to move towards the front, to which Brandt attaches great importance.

His "oblique lifting of the uterus" is noteworthy, because this form of lifting can hardly be accounted for in any other way than by the assumption that cicatricial connective tissue is to be stretched by means of it. Brandt* writes: "In case a lateral deviation of the uterus presents itself, we seek to execute the lifting in such a manner that the contracted parts are stretched, while the lengthened portions are made to contract." He then gives the advice, in case of extramedian uterus, to place one hand upon the contracted side—and only with this hand the lifting shall be executed—while the other hand exerts

*Brandt, Behandlung Weiblicher Geschlechts Krankheiten, Berlin, 1893.

a vibratory pressure, in order to *stretch* the *contracted parts* and
to stimulate the elongated ones to contract. And yet he per-
sists in his idea that he is using hygienic gymnastics for the uterine ligaments. So difficult is it for him to dismiss the erroneous
theory concerning the relaxation of the uterine ligaments.

All five forms of lifting have now been described in their
principal features. By means of each one of them the uterus is
displaced backwards and upwards, by the one last described any
laterally contracted tissue is stretched. In this way it is possible to force the uterus much higher up than we could do it
bimanually. The cervix escapes the finger in the vagina, and
stretching of the latter gives rise to a certain sucking sensation.

By means of these liftings Brandt is also curing a portion
of retroflexions, not all of them. In many cases he continued
the treatment for over a year, but most of those who are not
benefitted by it stay away after a few months, as he himself
declares, because their complaints have vanished.

In my clinic the liftings are still executed for the sake of
experiment, and their technique is demonstrated in the courses
of instruction. But not having as yet cured a single case by
means of liftings where other methods failed, I do not make any
more use of them in my private practice.

A case of retroflexion of movable uterus was cured by liftings, but at the end of four weeks relapse took place. With the
consent of the patient, twenty of my auditors undertook, one
after another, to replace the uterus, which times and again was
put in retroflexed position by me. The nineteenth and the
twentieth were unable to execute this manipulation, because it
was no longer possible to place the organ in retroflexion. The
displacement was and remained cured.

In about one-half my cases I effected a cure by massage
and stretching of the fixating bands. As soon as normal mobil-

ity is restored and abnormal fixations are gone, requiring two to
four weeks, we proceed to treat chronic endometritis, on which
occasion the cervix is finally, by means of a hook forceps, pushed
vigorously backwards and upwards and, if necessary, also laterally upwards. In many cases we are thus enabled to bring about
normal position.   In about one-half of the cases such a result
cannot be attained, although there are not wanting such in
which the uterus remains in anteflexion for three to eight days.
In such cases I assume that, owing to the continually abnormal
position, the peritoneum of the posterior, but above all of the
anterior, pelvic excavation has undergone such a change that
even a slight interference—overdistension of the bladder, coughing, sneezing, etc.—is apt to displace the uterus backwards
again.   Without any exception these are cases in which the displacement has, according to history, persisted for a long time.
Under such circumstances I introduce a pessary, forcing the
uterus into an anteflexed position for the time being, and giving
the peritoneum time to conform itself through the process of
involution with the normal position of the uterus.   In a majority of cases the pessary may be removed in two months, at most
three, and the displacement is permanently cured.

The difference between this and the ordinary pessary treatment lies in the fact that the pessary is endured without any
complaints, and is not a palliative, a kind of bandage, but leads
to a cure, an event of exceedingly rare occurrence unless massage and endometritis treatment have preceded it.   The most
favorable time for the involution of a faultily distributed pelvic
peritoneum is and still remains the time of involution in childbed.   The pessary is then introduced ten to fourteen days post
partum, and many a case of retroflexion is cured, even without
massage.   But we need not abandon all hope in patients not
confined to childbed.   Quite recently I observed a case, a per-

manent cure of which by means of the pessary was not effected
until two years after the treatment had begun. True, in such
cases the pessary may prove injurious to some extent, waste
material being apt to collect about it and to induce endometritis. This I endeavor to avoid by substituting, at the end of
eight weeks, a celluloid pessary for the rubber-covered copper
wire pessary. Most of the failures, however, are met with in
the most common form of deviation, in fixation anteriorly.

Those cases in which the contracting scar is situated in
the vaginal vault itself form a special group. In two cases of
this kind I effected a speedy cure, forceps delivery, from which
the injury resulted, having taken place respectively three and
four months before. In a third case a cure could not be effected
in spite of multiple transverse incisions with longitudinal sutures.
In a fourth case, in which I completely excised the two scars
and vigorously stretched the tissue under anæsthesia, subsequently
also preventing the recontraction of the operation scar by massage and stretching, a perfect cure was accomplished. The displacement had occurred two years previously. In general, the
same rule holds good in all deviations, more particularly in retrodeviations, nay, in every disease, that the prospect of an early
and complete cure is the more favorable the less time has
elapsed since its genesis. Not the age of the afflicted person,
but the age of the deviation aggravates the prognosis. This
proposition I can verify by a long series of cases. The results
achieved by Brandt's treatment will become still more marked
as soon as every physician has acquainted himself with it and
the most propitious time is chosen for its application. I cured
a case of retroflexion permanently in three sittings, a second one
in five sittings, and perhaps less would have been needed if I
had discontinued the treatment as soon as the position had
become normal. In that repeatedly mentioned case in which,

as it were, under my eyes, retroversion occurred owing to over-distension of the bladder, I replaced the uterus fourteen days post partum and introduced a pessary. The patient recovered. The only cases in which a cure by pessaries alone was effected were puerperæ. We need not therefore resort at once to massage in these latter cases.

I have not yet discussed the method of treatment and the results attained in fixed retroflexion.

In these cases, too, we must endeavor to replace the uterus as soon as possible, and as this can be accomplished only by severing the adhesions the latter manipulation is the first step in our procedure. It may be done in two ways, either by removing the adhesions by means of massage or loosening them under anæsthesia, according to Schultze. I must confess that the former was attended with success only in cases of not too long standing, or in such in which the area of the adhesions was not too extensive. It is possible, also, that I did not have patience enough. I tried to detach them by massage for four or five days each time, at times even for eight days. I tried the whole series of the above mentioned bimanual methods of replacement, and at the same time sought to dispel the adhesions by massage and to force the uterus away from the posterior pelvic wall. Although in many cases, at the beginning, only a very sensitive spot indicated the seat of such an adhesion, and the latter itself did not become visible until later, when the sensitiveness had decreased, I, neveerthless, succeeded not unfrequently in replacing the uterus. If, after four or five days, there should be no prospect of a cure to be attained by massage, I do not hesitate to loosen the uterus under anæsthesia, according to Schultze, the adhesions being severed with a firm hand, because the time of treatment is thereby considerably shortened. In using massage we fail to notice when the adhesions give way, for the reason, perhaps, that, by

and by, they become more succulent and softer and disappear
gradually; in applying Schultze's method, however, by which, as
a matter of course, possibly adherent ovaries are simultaneously
loosened, we feel distinctly when the adhesions part.   A sensation is produced not unlike that created by tearing gauze.   In
one case my assistants, too, could *hear* how the adhesion parted.
Even people who abhor the surgeon's knife, and generally dread
an operation, are easily persuaded to submit to such a treatment
on being told how much time and money could be saved by so
considerable an abridgment of the work.   In a few cases I did
not succeed until the second time, but never yet have I effected
a cure by means of massage alone after Schultze's method had
failed.   The latter happened to me even in a case in which, on
account of atypical hemorrhage, I had made a digital exploration
and was able to use the last mentioned intrauterine method of
Schultze.   Even on substituting the placental forceps for the
fingers engaged in intrauterine examination, and being able, by
virtue of the lever action of the long instrument, to exert such
a force that the muscular tissue of the uterus threatened to go
asunder, I had no better success.   Schultze himself reports a
case in the treatment of which he was unable to loosen the
uterus, but Brandt, to whom he sent the patient, succeeded in
accomplishing by massage what he (Schultze) had failed to
achieve by his own method.   I am, therefore, of the opinion
that in treating the above cases I did not have patience enough.
I never saw any harm accruing from Schultze's method, only
twice I noticed an insignificant effusion in the Douglas space.

The uterus having been loosened, I allow the patient to rest
for a day or two and I order an ice bag to be applied, in order
to influence the pelvic vessels by way of contraction.   As soon
as possible the treatment is to be continued, and if parametritic
bands are present, these real causes of retrodeviation are

removed by means of massage and stretching. Otherwise it is
at least expedient to massage sometimes the peritoneal surfaces
that have become agglutinated. The uterus being apt to relapse
in the beginning, I usually introduce a pessary during the anæsthesia, allowing it to stay for some time. Not long ago I was
even obliged, in two cases, to use a stem pessary, not being able
otherwise to keep the uterus in anteflexion, for the purpose of
preventing the two sore surfaces from coming into contact, the
latter being apt to reunite quickly like fresh wound surfaces.
Generally speaking, in cases of fixed retrodeviation the prognosis
concerning a permanent cure of the displacement is a very good
one, a majority of them finding permanent relief. In one case
the uterus even remained spontaneously in anteflexion as soon
as the adhesions had been loosened. However, this case is an
isolated one and dates from a recent period, not having had
observed one of this kind at the time I wrote my paper for the
Volkmann'sche Vorträge (1888). Nor have I asserted anywhere that the uterus remains anteflexed as soon as it is freed
from its peritonitic adhesions. On the contrary, from all this
we deduce the proposition that only after the removal of those
parametritic fixations which gave rise to the retrodeviation does
the uterus remain in normal position, unless the peritoneum of
the anterior and, for aught I care, posterior pelvic wall is over-
stretched or pulled out. It was Duehrssen who, in a meeting
of the Berlin Gynecological Society, combated the above asser-
tion, alleged to have been made by me.

In the few cases in which I did not succeed in loosening the
uterus, I massaged the adhesions, massaged and stretched the
parametritic bands, until the pains disappeared and mobility
was restored as much as possible, and then removed the chronic
endometritis. The symptoms disappeared in all, except sensitiveness on pressure and transient pains experienced during

greater exertions made in the locality where those adhesions had been. I did not therefore consider it advisable to sever the posterior peritonitic fixations by a life-endangering operation for the purpose of fastening the uterus to the anterior abdominal wall by means of anterior peritonitic fixations. The only thing to be gained from such a proceeding would be a transfer of the same symptoms from the posterior to the anterior region with the possible addition of bladder troubles.

One of the last mentioned patients called upon me a short time ago again on account of sterility. She stated that since her treatment (three years ago) she had been entirely free from any troubles. I suggested to her the less dangerous manipulation of my friend Boisleux, consisting in vaginal opening of the cul-de-sac and manual loosening of the adhesions by means of the finger introduced. But not being able to promise that the conditions of conception might be improved by the anterior position sought to be attained, and, moreover, at least a transitory disturbance of her physical welfare caused by peritonitic pains being almost unavoidable, the patient so far has failed to declare her willingness to undergo an operation.

Competitive Methods.

Gynecologists have long since been divided into two parties, each one holding a different view as to the mode of treatment. One of these started out with the assumption that the symptoms did not spring from retrodeviation as such, but from complicating affections, and consequently treated the pains only symptomatically with local and general narcotics, constipation with cathartics, etc., while the other party, on the contrary, looked to retrodeviation as the source of the troubles and regarded the replacement and anchoring of the uterus in normal position by means of a pessary as the only correct treatment. Of this latter party I was originally an adherent. It was the views of my teachers in gynecology that I followed. If at the present time I do not cling so tenaciously to their dogmas, it is because I have become acquainted with better methods to cure those complications, and have more frequently than ever been in a situation to remove all the troubles without simultaneous cure of the displacement. I should nevertheless feel aggrieved, if I had performed ovariotomy or castration without fastening the retrodeviated uterus to the abdominal wall. On the other hand, I know a great many women who, freed from all troubles, have been living contented for years, although the uterus is not in an anteflexed position. These would laugh at me, if I should suggest to them to have the displacement remedied by laparotomy. Such an attempt would be unwarranted because the intentionally produced traumatic peritonitis would provoke new evils. It is true, we are often compelled simply to replace the uterus, as has been the universal custom hitherto, and to support it by a pessary, because the patient coming perhaps from a distance has no time to spare for more thorough treatment. But in such a contingency we must not omit to call her attention to better therapeutic methods in case the result should not be satisfactory. In many cases the symptoms improve. We know, for instance, that in chronic parametritis and anteflexion support rendered by a pessary often removes the downward pressure and pains. I am convinced, too, that the frequent bimanual replacing in selecting and adjusting a pessary is equivalent to a massage cure on a small scale. For this reason the more difficult the adjustment of the pessary was, the more complaints disappeared. The Hodge pessary, the Schultze S-pessary, the Thomas pessary, and my fork-pessary are good pessaries. All may be bent and formed by the physician himself from celluloid rings according to Schultze's direction. Cleanliness attained by slightly antiseptic douches is imperative in using the pessaries, as otherwise decubitus and endometritus will ensue.

Suturing to the anterior abdominal wall, ventrofixation, was once proposed by P. Mueller and Koeberle and executed by Olshausen. The improvements made by Saenger, Leopold, Fritsch and others, refer only to the manner of fastening. My own experience in this direction is of an unfavorable character. In only two cases I operated—in one of them the situation was aggravated by a relapse of double ovaritis, in the other there was a solid intraligamentary ovarian tumor. The latter case I lost, the patient dying from pernicious vomiting; the former is still living, but how? At the time being the uterus lies in the best of anteflexion movable over the bladder, united with the anterior abdominal wall only by peritonitic adhesions. The menstrual bleedings, in spite of the removal of both cystically degenerated ovaries, have become more frequent, more painful, and of longer duration, and besides there is occasionally severe vesical tenesmus. I shall next devastate the mucous membrane of the uterus, according to Dumontpallier, by means of zinc chloride, two years and a half having elapsed since the operation without any satisfactory result. The poor woman (C. R., of F.-B.), under proper treatment, had been repeatedly free from all troubles. Often the uterus remained for eight days spontaneously in anteflexion. Pessaries could not at all be endured on account of the sensitiveness of the vagina. It was at that time that Saenger, during a meeting of the Congress of German Naturalists and Physicians, at Halle, in 1891, had presented cases of vagino-fixation. Saenger introduced the needle on the posterior lip of the cervix, ran it through the uterine tissue up to the insertion of the Douglas fold, bringing it out and back again into the vagina by way of the vaginal vault.

This could be done only if the uterus lay in a retroflexed position, the vaginal portion of the cervix was drawn upwards, and
the vaginal vault pressed down. After two or three such sutures
had been introduced the uterus was replaced, the sutures tied,
and a tampon placed in the vagina for several days.

To me it was clear that Saenger's results were achieved
only through provocation of an adhesive peritonitis between the
cervix and posterior pelvic wall under elevation of the former
by tamponing. I therefore proceeded in the following way
with my patient in question: Introducing a needle horizontally
in the posterior vaginal fornix and running it through the tissue
of the cervix, with thumb in the vagina, index finger in the rectum, I drew a portion of the latter down from above as high as
I could reach, brought the vaginal fornix and rectum into contact with each other, and now, introducing the needle near the
point of exit and running it out near the point of ingress of the
first stitch, I carried it through the anterior wall of the rectum,
without it appearing in the lumen. This carrying of the needle,
for a space of 2 cm. width, through the wall of the rectum
without the suture lying inside of the rectum, was easier to
accomplish than I imagined. The suture was not tied in the
vaginal fornix until, the uterus being drawn down as much as
possible and the bladder pushed aside, a second suture was
introduced about in the middle of the anterior vaginal wall and
carried in such a manner through the tissue of the uterus
pressed up by a strong sound, that the corpus of the uterus was
included in the stitch and the latter brought out again at about
the middle of the anterior vaginal wall, near the entrance
stitch. The second suture was first tied and then the first.
Tampons were then introduced into the vagina for several days.
The sutures, being catgut, were not removed, and the patient
was dismissed after two weeks with anteflexed uterus. At the

end of two more weeks I was hastily summoned, the patient
suffering from vesical tenesmus and pains so severe as to
nearly take her breath away.   I found the uterus retroverted
and anteflexed.   Upon replacement the symptoms disappeared.
In a few days she had a renewed attack and so forth, as
soon as she arose from bed.   Finally, I was obliged to remove
the artificially produced adhesions by means of massage, in
order to allay the pains.   Once more I dismissed the patient
free from trouble with retroflexed uterus.   In the summer of
1892, however, she had another relapse.   I then undertook
castration and ventrofixation, with the above named unfavorable result.   For the benefit of the sufferer I wished I had
attempted total extirpation and vaginal castration.

This case proves how correctly taken the standpoint is
when Fritsch remarks that operative treatment of retroflexion
is the one for the poorer class of people.   If my patient had
not been obliged to work, the relapses would not have followed
each other so quickly and finally would have ceased altogether,
apart from the fact that a pessary could have been worn.   As
operations cannot be performed so easily in the outside practice
as in clinics, it is preeminently theirs to handle such cases.

In the autumn of 1892 I operated on a second case according to the mode of vaginofixation contrived by me (Mrs. L. of
P.).   In this case the uterus remained permanently in anteflexion,
but with the result that the patient, who had previously been
free from ailments in spite of retroflexion, complained now, with
the uterus in its right position, of pains in the abdomen and constipation.   By means of massage I lessened the latter, but I did
not dare to continue it for fear the adhesions holding the uterus
might disappear along with the pains.

Since that time I have made no more attempts in this
direction, nor did I feel justified to raise any hopes after the

improved forms of vaginofixation of Mackenrodt and Duehrssen had become known.

On the same occasion, when Saenger read his paper in Halle, Schuecking's instrument was exhibited by Professor Schwartz—a tube in form of a sound, armed inside with a handled needle that can be pushed forward, thereby describing something of a semi-circle. The uterus was firmly drawn downwards by Schuecking, the bladder pushed aside by an assistant, the sound introduced into the uterus, the needle pushed forward so that it came out at the anterior vaginal wall. The needle was now armed with thread, drawn back, and the suture tied. This suture, if correctly applied, was to run through the cavum uteri, fundus, peritoneum of the vesico-uterine excavation, and to reappear behind the bladder through the anterior vaginal wall. This proceeding was soon combated by Saenger, improved by Zweifel, and permanently banished by Fritsch and his assistant, Glaeser, who demonstrated that the bladder was injured thereby in a majority of cases and the ureter occasionally. And yet it was the prototype of the modern vaginofixations.

Mackenrodt published his method first, in connection with a case of anterior colporrhaphy. It consists in longitudinal incision upon the anterior vaginal wall, separation and lateral dissection of the flaps, blunt loosening of the bladder from the uterus, drawing down of the latter until the fundus and both wound margins can be reached with one suture; hence all backwards running sutures, which pass through both wound margins and take up the anterior wall of the uterus, so that finally we have a sutured wound in the anterior vaginal wall, to which, the bladder being displaced forwards, the anterior surface of the uterus is stitched on.

Duehrssen's operation is based on Zweifel's improvement of Schuecking's proceeding —transverse incision in the anterior

vaginal fornix, loosening of the uterus from the bladder, suturing
of the corpus uteri to the anterior margin of the vaginal wound
by a row of transverse sutures, suturing of the vaginal wound.
Duehrssen formerly used fil de Florence, but as the sutures had
in many cases been pushed off into the bladder, thereby causing cystitis and calculi, he began recently to employ catgut. Not
until after the operation does Duehrssen use massage in cases
in which complaints are present. In this way there is no such
a thing as a cure of retroflexion without an operation.

The author has, for some years past, hit upon a method
closely related to that of Zweifel—transverse incision in the
anterior vaginal fornix, loosening of the bladder, drawing down
of the peritoneum of the vesico-uterine excavation until it
tightens, then suturing of the fundus to the peritoneum by a row
of transverse sutures. Zweifel has done exactly the same thing,
but under laparotomy. It is not absolutely necessary to remove
the quilt-like sutured pouch of the peritoneum, but if it should
prove an obstacle there is nothing in the way of our doing so.

The condition thus created, defect of the vesico-uterine
excavation, may be congenital. I have observed and illustrated
such a case (see Fig. 15).* In those apparently numerous
cases disclosing a peritoneum of the vesico-uterine excavation
grown to abnormal length and in those exhibiting it as having
become pathologically adherent to the cervix, this therapeutic
method is etiological. It brings together tissues which, under
physiological circumstances, are also in close proximity; it
creates a condition occurring physiologically, and does not join
organs that do not belong together, as, for instance, the fundus
uteri and vagina. Moreover it does not constrict the bladder,
as is the case, according to the investigations of Mackenrodt,

---

*Compare also On Physiological and Pathological Adhesions of the
Uterus. Plate I, Fig. 7.

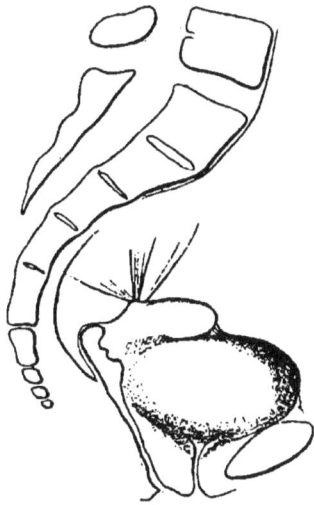

Fig. 15. Congenital Defect of the vesico-uterine excavation in a child.

with the two last mentioned methods. Besides, the uterus is elevated by this method, while it is drawn downwards by the latter. It is years since I have planned this method, but being candid enough in case of patients who were not cured of the displacement by means of massage and wearing a pessary, to set forth to them the chances of a suggested operation, namely, that those pains which they are at times feeling as caused by chafing fundus in the rear, would certainly be experienced for some time at the front, no one has as yet been found willing to submit to such on operation. Peritonitic, perhaps also bladder troubles, even if only transitory, are sure to result also from this operation. Possibly the latter might be combined with a second operation, consisting in opening the cul-de-sac from the posterior vaginal fornix and suturing of the visceral peritoneum to that of the posterior wall of the pelvis. To do the latter we of course would have to draw down the peritoneum of the posterior pelvic wall until it tightens.

The question now is, "What is accomplished by all the other artful operations, if at the end the uterus is forced into an anteflexed position, regardless of any complicating inflammations?" I claim, nothing, unless it be the addition of troubles engendered by traumatic peritonitis. These complaints do not result from retrodeviation, but from the complications. They are found present in the complicating diseases just as well, without retrodeviation; and will persist in spite of oderative replacement, with the exception of a few pressure symptoms.

12

PROLAPSE.

The curing of prolapse constitutes the weakest part of the whole of Brandt's therapeutic method, and Brandt himself told me once that if he were asked to give up a part of his method he would most willingly abandon his treatment of prolapse. On one hand, Brandt himself estimates that he had success in only seventy per cent. of the cases treated, including even improved cases, relative successes and relapses; on the other hand, there are just in this direction so many excellent competitive methods as can be found in no other branch of the treatment. If this is admitted by Brandt himself, who, as the inventor, employs the procedure with an entirely different love for the thing, with entirely different skill and endurance; who, in spite of his reputation, charges smaller fees than most of his imitators,* and for this reason can keep his patients longer, he is perfectly right in declaring it as a mistake made on the part of his friends when they emphasized his curing of prolapse as the particularly striking thing in the first of his publications on his modus operandi.

Yet no case should be treated in another manner before Brandt's method has been given a trial, because a permanent benefit is always secured even if the prolapse itself remains uncured. It is the unanimous sentiment of the most experienced gynecologists, corroborated by a large part of the cases examined by me with this particular object in view, that but few prolapses occur without inflammations being found in the annexa.

---

*This question of fees is mostly passed over in silence in medical publications, and yet it is a matter of importance whenever we calculate the value as compared with that of competitive or rival methods. Brandt charges now 100 crowns (about $26.00) per month, formerly only 50.

These affections of the annexa in case of prolapse produce, as a
matter of course, the same complaints as without it; but as long
as the prolapse lasts they are, as it were, deadened by the complaints engendered by the prolapse and do not reappear until the
latter is cured.   The removal of these inflammations, too, is
sure to be accomplished by a trial of Brandt's therapeutics as a
side effect.

To be sure, it would be better if, in diagnosing a case, we
were enabled from the beginning to say whether prolapse can
be treated according to Brandt's proceeding, or whether it must
be operated.   As I already suggested on page 89 of my paper
in Volkmann's Vortræge, there is a certain contrast between the
two therapeutic methods, in so far as those very cases that cannot be cured by means of massage may, with an appropriate
method, be easily cured by an operation.   However, I am not
yet ready to pronounce a final and definitive judgment, and am
therefore compelled to enter more fully and accurately upon the
definition and etiology of the displacement, in order to be more
thoroughly understood.

DEFINITION.

A prolapse, according to a layman's view, is everything
that protrudes from the openings of the body, but which apparently belongs into it.   With regard to prolapses of the female
genitals, this is, even scientifically considered, not much different, for even if tumors of the uterus in their spontaneous elimi-
nation are apt to be confounded with prolapse, they often drag
the pelvic organs down with them and produce a real prolapse.
Quite recently I operated on a case in which a large vaginal
cyst had given rise to cystocolpocele and dragged the uterus
downwards.   A second case I operated on several years ago.
According to the organs involved in the prolapse, we distinguish
(1) cystocolpocele and (2) proctocolpocele, both with or with-

out descent of the uterus. The latter is the case when the external os of the uterus has not yet proceeded beyond the entrance of the vagina; but if we find the cervical opening at the prolapse, the condition thus created is a prolapse of the uterus.

There are, moreover: (1) Total prolapse of the uterus; (2) partial prolapse of the uterus, according as the fundus uteri has left the vulva or not. The former is rarely met with, the latter seldom without inversion of the vagina, seldom without cysto-colpocele or proctocolpocele or both. We further distinguish prolapse caused by hypertrophy, (1) of the vaginal portion of the cervix, (2) of the portio intermedia, and (3) of the portio supravaginalis. The fundus uteri may nevertheless be found in its right place in the pelvis.

ETIOLOGY.

To a large number of theories about the etiology corresponds a much larger number of curative methods, which latter possess to a great extent merely a palliative value.

1. Huguier[*] found that in all cases of prolapse the cervix was hypertrophied and elongated. Quite consistently he sought to effect a cure by amputation of the vaginal or even supravaginal portion of the cervix. Cystocolpocele he regarded as the consequence of cervical hypertrophy.

2. Marion Sims[**] on the contrary, in view of the successive phases of prolapse recurring after the replacement, arrived at the conclusion that cystocopocele occurred first, which in turn drew down the uterus and then the posterior vaginal wall. A wide pubic arch, in his opinion, is a predisposing element. His therapeutics, accordingly, was directed first to the anterior vaginal wall.

---

[*]P. C. Hugier, Mémoire sur les allongements hypertrophiques du col de l'utérus, etc., Paris, 1860.

[**]J. Marion Sims, Klinik der Gebaermutterchirurgie, Deutsch von Beigel. Erlangen, 1865.

3. Schultze* holds the relaxation of the Douglas folds responsible for a majority of cases. In the rich, who can afford to take good care of themselves, retroflexion only arises, while in the poor, compelled to work, stand, walk, and exert their abdominal muscles, both retroflexion and prolapse are found.

A relationship between retroversion and prolapse was pointed out by Sims already (l. c., p. 233): " In case of anteverted uterus a prolapse is perfectly impossible, no matter what the accompanying circumstances may be."

In conformity with the facts Schultze reports that in a great many cases the uterus, after the replacement of the prolapse, lies anteriorly. Defects of the perineum and incomplete involution of the hypertrophy of the vagina incurred during pregnancy are predisposing elements.

4. In Von Winckel's book ** we find related a case of congenital prolapse (hypertrophy of the vaginal portion) in a child that was carried up to time and had spina bifida. Trophic nerves seem to have exerted their influence in this case.

5. Fehling,*** besides the causes mentioned hitherto, is of the opinion that retrodeviation is not the cause, but the consequence of prolapse. He likewise speaks of an over-distended bladder as influencing the genesis of cystocolpocele.

6. According to Schroeder's **** theory total prolapse results from partial prolapse, because the hypertrophied uterus diminishes in size while the os uteri remains unaffected.

7. Kuestner***** measured the intraabdominal pressure and found it to be 40 cm. in upright position, 15 cm. in recum-

* B. S. Shultze, Lageveraenderungen der Gebaermutter, Berlin, 1881.
**Von Winckel, Lehrbuch der Frauenkrankheiten. Leipzig, 1890.
***Fehling, Lehrbuch der Frauenkrankheiten. Seite, 269.
****Schroeder, Lehrbuch der Gynaekologie.
*****Kuestner, Grundzuege der Gynaekologie. Jena, 1893.

bent position, and negative in knee-chest posture. The effect
of this pressure upon the anterior wall of the uterus in retrodeviation is, according to his opinion, the most frequent cause;
partial defect of the perineum a predisposition.

Total prolapse he believes to be the consequence of involution processes of the uterus that had been hypertrophied,
whereby the latter decreases in size from the fundus towards
the external os. The hypertrophy is mainly of an œdematous
nature, produced by the difference in pressure between the
atmosphere and the abdomen.

Fritsch* emphasizes the difference between secondary prolapse resulting from cystocele, and primary prolapse, in which
the uterus sinks down first and simultaneously inverts the whole
circumference of the vaginal vault.

This selection of the most varied views concerning the
genesis of prolapses might be amplified indefinitely. Of the
most recent authors, Zweifel, Martin, Heitzmann, Herzfeld, and
others have not yet been heard, and there would be no end of
opinions if I were to quote from older writers and authors of
foreign countries. Only one theory yet deserves to be mentioned, because it disregards mechanical principles entirely,
trying to account for everything by nervous causes, by paralysis
or partial paralysis (paresis) of nerves and centers. This is the
theory of Dr. Johannes Hyri.** In reply to it I quote the words
of Lichtenstein:

"It is much more exasperating to account for a phenomenon by a little mechanics (mechanism) and a powerful dose of
the supernatural than merely by mechanics." For me there has
ever been something supernatural in every attempt at accounting for certain phenomena by nervous influence with its paral-

*Fritsch, Die Krankheiten der Frauen, Berlin, 1894.
**Dr. Johannes Hyri, Centralblatt fuer Gynækologie, 1894, No. 2.

yses, pareses and centers. Therapeutics has, in spite of electricity, not achieved any surprisingly great success. If mechanical principles enable us to give satisfactory explanations, the
results of mechanical therapeutics will corroborate their correctness.

In order to ascertain whether the above mentioned theories
are correct or not, I extended my repeatedly mentioned investigations to an examination of the folds of Douglas and
the pelvic floor. From the very outset it was my intention to
prove the causal value of the folds of Douglas for the anteflexed position of the uterus. In the course of my experiments,
however, I arrived at the conclusion that, in case everything is
in a healthy condition, none of the so-called ligaments of the
uterus is strained or tightened for the purpose of keeping it in
its normal position, and just when all the tissues of the uterus
and its surroundings are unchanged, the mobility of the organ is
greatest.

What usually secures for the uterus its position in the pelvis
is the form of the pelvic floor. The main part of it is formed
by the levator ani. This muscle does not represent a funnel—
this shape being obtained only by anatomical preparation--but
it forms a groove or channel (fossa). The punctum fixum, or
fixed point, of the muscle is the anterior semicircle of the pelvis,
up to the spines of the ischium; the punctum mobile, or movable
point, is its raphe behind the rectum and the coccyx. This
muscle is covered internally and externally with a fascia, and
strengthened by the sphincter muscles of the anus and bladder,
the ischio-cavernosus sive constrictor cunni, and the transverse
muscles of the perineum. All these, physiologically, form one
muscle, as none of them can be contracted singly; anatomically,
they are merged into each other by muscular prolongations. The
rectum, vagina, and urethra, which pass through this muscle,

are surrounded by its fibers.   I believe that the transversus
perinei profundus is nothing else but muscular fasciculi, passing
from the levator ani to the sphincter vesicæ and being analogous
to those running from the levator ani to the sphinctor ani.

Contraction of this muscle causes: (1) Narrowing of all
openings situated at the pelvic outlet; (2) leveling down of
the fossa until almost a plane is formed sloping from the
front towards the back; (3) increase of the angle opening
towards the rear and formed by the three tubes (rectum, vagina,
and urethra) in passing through this muscle, and approach of its
apex to the symphysis.

The apertures in the pelvic floor for the passage of these
three tubes are situated 6–7 cm. in front of the spinal line (see
Fig. 16), at the spot where the vaginal portion of the cervix is
usually found.   It follows that the vagina, from its passage
through the pelvic floor up to the vaginal fornix (a space 6–7
cm. wide), runs almost horizontally backwards.   The lower seg-
ments of these three tubes are intimately blended with each
other; behind the muscular fasciculi of the folds of Douglas
form the last spurs, which, however, do not reach the rectum;
in front, almost as far as the internal os, the uterus is connected
with the bladder by muscular bundles (see Fig. 7, 8 and 9).

The very manner in which these organs are grown together,
as well as the form of the pelvis, force the organs into such a
position that the uterus is anteflexed.   The least resistance to a
displacement of the uterus is offered perhaps by the pelvic
peritoneum; because all around the lower portion of the abdominal cavity it is movable in a high degree.   Its distribution over
the pelvic organs is such that it can be best illustrated by assuming that a peritoneal funnel once intruded, suspended between
uterus and rectum, into the pelvis, and that, from the front and
below, the inner genitals and the bladder were pushed upwards,

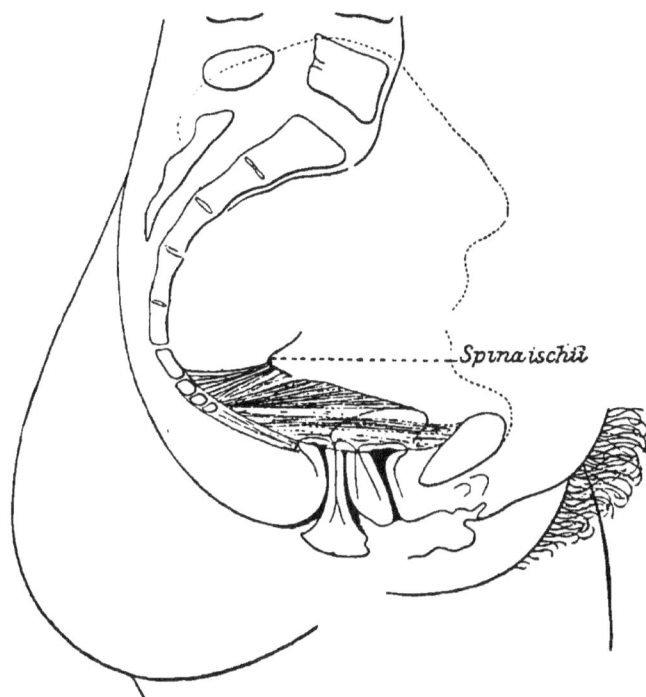

Fig. 16. Pelvic floor and levator ani shematically in order to show its form
and the distance of the passage of vagina through it from the spinal line (where
the vaginal portion of cervix is situated.) ⅓ nat. size.

where finally, by means of muscular prolongations from the uterine muscles, they became firmly grounded. In front the peritoneum is movable up to the umbilicus; in the rear, a traction on it is propagated to the diaphragm; only over the promontorium it adheres more firmly to the substratum.

These anatomical facts premised, I shall briefly relate the results of my experiments made at the time. We read l. c. page 28 :

"Investigations concerning the pelvic floor I instituted only in about the last fourth of the cases examined by me, its great importance not having become clear to me until later.

"The uterus, when forced downwards vertically or in the direction of the pelvic axis, is, if antiflexed, pressed only more firmly against the pelvic floor, and its position and attitude are so much better secured in case we leave the vaginal portion of the cervix in the spinal line. Like a bolt, it is placed transversely before the opening in the levator ani muscle. But if we bring the vaginal portion forwards beyond the opening of the vagina through the levator ani, and press the uterus downwards in the direction of the vagina as it is in retroverted position, the vaginal portion of the cervix may be pressed downwards several centimeters into the part of the vagina situated below this opening. The vagina becomes thereby inverted. If now we sever the peritoneum at the pelvic entrance, the uterus does not by any means become more movable, nor does the blunt separation of the pelvic connective tissue have any effect."

Not until the vascular bands at the sacro-iliac articulation and the sacral nerves are severed, does the uterus, especially towards the front, become a good deal more movable and can be pressed downwards into the portion situated below the lavator ani muscle. By this manipulation the fixations of the

vagina to the rectum, and especially those to the bladder, are
greatly stretched.

When released, the fundus uteri glides back spontaneously
and with a peculiar noise, upwards beyond the levator ani.   In
spite of all my efforts I was not able to move the fundus of the
normal uterus beyond a point corresponding to the plane of the
levator ani muscle.   Even on a completely excised specimen I
could not force the uterus any farther.   In this case the vaginal
portion of the vervix protrudes, 2 to 3 centimeters wide, from the
entrance of the vagina, the external os being directed forwards
and upwards, while in spontaneously prolapsed uterus it is well
nigh always turned backwards and downwards.   (Comp. Klob,
Pathological Anatomy of the Female Genital Organs.)   It seems,
accordingly, as if the fixations extending from the uterus to the
bladder, or rather to the symphysis, opposed an especial resistance to the protrusion of the uterus, by keeping the cervix drawn
tightly forwards and upwards.

It may be mentioned, by the way, that these forcible inver-
sions of the uterus into the lumen of the vagina frequently produce lacerations in the vaginal walls, these rents having been
found regularly in the lateral walls.   It seems that the looser
fastening of the anterior and posterior vaginal walls (of the
columna rugarum) enables them to be stretched to a greater
extent than the more thoroughly fixed lateral vaginal wall.   This
may account perhaps for the two-pointed shape of perineal lacerations as soon as they extend up to the region of the columna
rugarum.

Having discussed the fixating properties of the pelvic floor,
together with those of the vessels and a portion of the fixations
to the neighboring organs, I shall revert to those of the pelvic
peritoneum.

If the severing of the whole pelvic peritoneum, as stated

above, has no visible effect on the mobility of the uterus, it is clear that a severing of a part of it, for instance of the folds of Douglas, will likewise have no influence. If now Fritsch (1. c.) restricts the results of similar experiments made by Kirvisch, Savage, and others in so far that a part of the effect exercised by the folds of Douglas is to be ascribed to the underlying vessels, but remarks that the severing of the peritoneum is impossible without partial cutting of the vessels, I can most positively state that to these vessels and to the sacral nerves is almost exclusively due the backward fixating effect. I have no difficulty in accounting for the results attained by the above named authors, if I assume that, along with the peritoneum, they also cut the underlying vessels and nerves totally or partly."

Accordingly, it is not the folds of Douglas, nor, as I believe, the vessels and nerves, which keeps the uterus in its position. If it were otherwise, in gynecological operations, during which the uterus is forcibly drawn down and frequently kept outside for hours, they would often have to be stretched to such an extent as to cause prolapse, a contingency which, to my knowledge, has never yet occurred.*

But before it is possible to press the uterus forwards, the vaginal portion of the cervix must be brought forward to the opening in the levator ani muscle or the opening in the levator must be brought back to the cervix.

The conditions favoring the genesis of prolapse are present when the vaginal portion of the cervix is in front of the vaginal opening of the levator ani muscle, i. e., of the pelvic floor.

Gravitation and intraabdominal pressure are the forces pushing it forward.

---

* On the contrary the favorable effect exerted by such operations on chronic inflammations must be accounted for by the assumption that pseudo-membranes and parametric adhesions have been torn by stretching.

The pressure prevailing in the vaginal vault * is equal to that in the abdominal cavity; lower down it cannot be measured, the instruments being forced out or muscular contractions affecting the measurement. During walking and standing the atmospheric pressure is less than that in the abdomial cavity, and has, therefore, a sucking effect, similar to that of a cupping glass, and the conditions resemble those which, during birth, lead to the formation of a caput succedaneum.

This theory, based on anatomical studies and careful consideration of facts, I set forth as early as 1887, during a meeting of the Munich Gynecological Society, and it may be found in my contribution to Volkmann's Vortræge.

Let me quote from it, p. 39, ff.:

" *The conditions favorable for the genesis of prolapse of the uterus are present as soon as the cervix is in front of the opening of the levator ani, becoming exposed to the effect produced by the difference in pressure between the atmosphere and intraabdominal pressure.* This condition may be brought about: (1) By shrinkage of the pelvic connective tissue, dragging the vaginal portion of the cervix in front of this opening; (2) by relaxation of the levator ani, or rather incomplete involution of its opening after a birth, so that this opening is wider and extends further backwards.

"Both conditions compete with each other to a certain extent. If the opening in the levator ani muscle is large and the shape of the latter is funnel-like in consequence of relaxation**, no fixation, or at least only a slight one, of the cervix towards the front is needed, in order to bring the vaginal portion of the cervix under the influence of atmospheric suction.

* Both Schwartz and Wiedow concurrently stated this during the meeting of the Congress of Gynecologists in Halle on Whitsuntide, 1888.

**Brought about by senile atrophy, prolonged attacks of cough, influenza, excessive abdominal pressure, repeated lifting of heavy burdens, etc.

If the pelvic floor is tense and the opening narrow, as is the case when no birth has preceded, a very short fixation is needed in order to expose the vaginal portion of the cervix to the difference in pressure between the atmosphere and intraabdominal pressure.

"The intraabdominal pressure is greater than that of the atmosphere as long as the patient is standing, sitting or cowering, the atmosphere sucking, like a cupping-glass, on the vaginal portion of the cervix, much more so in proportion as the intraabdominal pressure is augmented by the abdominal pressure. The first visible effect of this is œdema of the cervix, liable to be increased by any disturbance of circulation due to mechanical constriction. Moreover, according to Beyer's investigation, the muscular fasciculi of the uterus are drawn out gradually more and more, producing a thickening and enlargement of the vaginal portion of the cervix, together with a thinning and lengthening of the cervix. This attenuation of the supravaginal portion of the cervix may assume such proportions as to destroy the continuity between the cervix and corpus (Klob).

" We can fully understand now the frequent co-incidence of prolapse and retroflexion, fixation of the vaginal portion of the cervix anteriorly being the etiological factor for both simultaneously. Equally easy to account for are those cases in which polyps situated in the vagina produced prolapse, which latter disappeared after their extirpation. The puzzling case related by Hohl, in which a girl fourteen years of age sustained a prolapse which disappeared spontaneously on her reaching the age of full puberty, is explained very simply by the increase of the pelvic incline incidental to her sexual maturity. Likewise it is easy to understand why at the beginning prolapses recede spontaneously in recumbent and side position (the intraabdominal pressure becoming less), but reappear in standing posture.

"As a special form of fixation anteriorly we have to set down those cases in which a cystocolpocele draws the vaginal portion of the cervix forward, as first observed very accurately by Marion Sims. In such cases the uterus is in reality more movable than when normal, the vaginal portion of the uterus, not unfrequently even in anteflexed position, being perceptibly drawn forward by the rapidly protruding cystocolpocele during any attempt made to increase the abdominal pressure. These are precisely the cases in which the support of the pelvic floor is wanting and the opening in the levator ani muscle is too large. If such patients are asked whether a full or an empty bladder affects the relative extent of the prolapse, they state positively that a full bladder tends to lessen the affliction. Found in a similar condition were the cases examined by me, in which, according to all appearances, there existed prolapse during lifetime (see Fig. 17). In these cases the peritoneum was found to be adherent to the apex of the bladder, preventing the latter's distension upwards. For this reason the bladder was obliged, on being distended, to expand towards the vagina, and by carrying the vaginal wall through the opening in the levator ani muscle in front of itself, led to cystocolpocele, and, finally, to prolapse of the uterus. The bladder, on filling up, always tends to expand in the direction offering the least resistance. If it encounters less resistance below, it will naturally expand in this direction. In both cases this process had been checked by a pessary, and, as a consequence, the bladder had formed diverticulum-like excavations under the broad ligament.

"I do not believe that a single prolapse is due to relaxation of the ligaments, as generally supposed up to the present time. This view is also contradicted, among other cases, by a case from the women's clinic in Munich, which I was enabled to

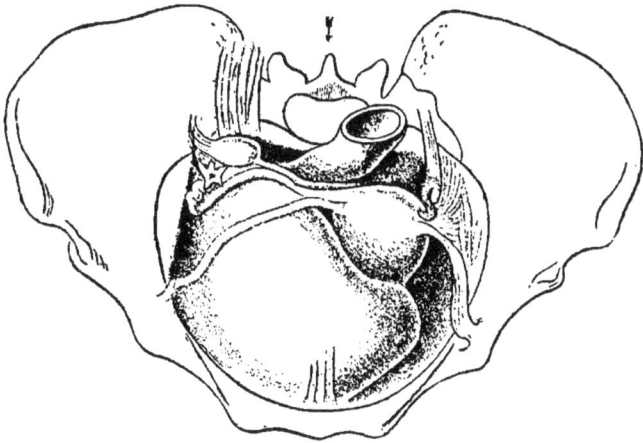

Fig. 17. Altered form of the bladder in case of prolapse of the uterus. Peritoneum adherent to bladder, that of the right broad ligament undermined, in autopsy, ⅓ natural size.

examine anatomically, and which exhibited the muscular fibres*
extending from the uterus to the peritoneum of the folds of
Douglas and of the whole broad ligament as thickened in a high
degree. Conformably with this, Kuestner already had observed
that, in retrodeviation, the muscles of the folds of Douglas were
not atrophied, but hypertrophied on account of labor hypertrophy."

The correctness of this theory has been practically confirmed by the experience gained ever since it was propounded.
Nor does it, strictly speaking, oppose any of the above quoted
theories, but rather complete them. Thus, for instance, the
theory of Huguier fails to assign any reason why the cervix
became hypertrophied, what induced it to become such, whereas
my theory accounts for it by the difference in pressure between
atmosphere and abdomen. Beyond the opening in the pelvic
floor the corpus uteri, forming, with all its annexa, a mass
of vast proportions, is held back by its volumen, or rather by
the muscular tissue of the so-called ligaments, while below the
atmosphere, as it were, is sucking at the vaginal portion of the
cervix or even at the intermediate portion, that is to say, in case
the anterior wall of the vagina has, as it often happens, been
drawn down together with the bladder. The next consequence
is œdematous infiltration and congestion, affecting all parts lying
below the levator ani, and inducing the covering epithelia to
become swollen and subsequently hypertrophied. The wandering cells of the œdema are changed into cicatricial tissue; besides,
every available fibre of the muscular tissue of the cervix is drawn
out, and the muscular fibres become elongated and increase.
Only hyperplasia and hypertrophy can account for those degrees
of increased volume that are met with sometimes. That part of

* The muscular tissue of the round ligament, too, was hypertrophied, and
drawn out to such an extent that the insertion of the tubes had been removed
over 1 cm. towards the median line.

the uterus which lies between the congested region and the corpus uteri, the supravaginal portion, becomes attenuated, but only passively lengthened. For this reason we speak of it only as of an elongatio colli. That the supravaginal portion of the cervix is not hypertrophied, but merely stretched, has already been pointed out by Marion Sims (l. c. p. 236); according to Klob (see above) it may tear asunder. For my part, I never found it thickened. The idea of stretching is materially favored by the fact that the cavum uteri becomes almost instantly measurably shorter as soon as the prolapse is replaced.

What might have been mentioned before, namely Schroeder's division of the cervix uteri into three sections, is by way of supplement, set forth in this connection. This division is important in so far as all three, each one by itself, may become hypertrophied, or, rather, the third one only elongated. By drawing a straight line from the anterior vaginal fornix towards the rear we bound the vaginal portion of the cervix above. If now we draw a straight line from the posterior vaginal fornix towards the front, we have, enclosed between these two lines, a portion one-half of which lies in the vagina, the other half beyond it, the attachment of the vaginal fornix extending obliquely from backwards and upwards towards the front and downwards. This part is the intermediate portion of the cervix, unknown to older writers. The remainder of the cervix, from the posterior vaginal fornix to the internal os, is the supra-vaginal portion.

In case the vaginal portion is hypertrophied, both vaginal fornices are at an approximately normal height; if the intermediate portion is hypertrophied, only the posterior, the anterior fornix being transformed and inverted. In case the supravaginal portion is elongated, the posterior fornix is also inverted and lower down.

A differential diagnosis between these two latter conditions can often be made only by examining the posterior covering of the cervix as to whether it is movable or firmly lodged. If fixed, we have a case of hypertrophy of the intermediate portion; if movable, it is an elongation of the supravaginal portion.

Mixed forms of the last two types, namely such as are adherent for a certain distance, but movable higher up, are the forms most frequently met with.

Continuing to discuss the afore-mentioned theories, the one of Marion Sims is the first to be approached. This theory, too, which, to be sure, fits a majority of cases, and according to which a cystocolpocele first draws down the uterus and only subsequently the posterior vaginal vault, fails to inform us about an interesting point, namely, what may have been the cause of the cystocolpocele.

Besides lacerations we must take into consideration disturbances of the involution of the opening in the pelvic floor so vastly enlarged by the descent of the child intra partum,—disturbances caused by too early leaving of childbed, lifting of the baby, excessive distension of the bladder, stretching of the os during therapeutical manipulations, or by abdominal straining in coughing during the puerperal period. Most of the prolapses occur in childbed. Incomplete involution of hypertrophy of the anterior vaginal wall formed during pregnancy, operates as an additional predisposing element, and is present mostly in the same cases. But as to the perineal defect, also the partial one, my experience does not justify me in attaching so great a significance to it. The very neck of the bladder is so firmly fixed under the symphysis by the middle ligament of the bladder and both lateral ligaments, that I am inclined to believe the urethra is only secondarily turned inside out on account of the cystocolpocele occurring more in the middle part of the anterior vaginal

13

wall. Moreover, there are so many perineal defects without
prolapse and so many prolapses with an intact frenulum, that I
firmly believe it cannot be found more frequently in prolapses
than otherwise in women of such an age.

Into this category belong a number of cases in which posterior or lateral parametritis had existed with anteflexed position of the uterus, and where, with the removal of parametritis
by means of massage and stretching the cystocolpocele also
yielded. For these cases I can offer no other explanation than
the assumption that parametritis had begun to fixate the uterus
downwards. The bladder would undoubtedly have displaced
the uterus backwards and retrodeviation would have ensued, if
coughing or other exertions of abdominal pressure had not
obliged the bladder to turn downwards. These were for the
most part cases in which the cystocolpocele protruded sufficiently only when the patient was in an upright position. They
were all cured, one excepted, who had a relapse—an unmistakable indication that parametritis had been the cause. Here
Fehling's etiology finds its verification.

In cases of prolapse resulting from cystocolpocele, only the
anterior vaginal wall is completely inverted. We also find in
most cases that the posterior vaginal fornix has descended
deeper, but there is still extant a portion of the vagina 5–7 cm.
long. The rectum is in its proper place. Proctocolpocele and
cystocolpocele simultaneously existing I have met hitherto only
in a few cases of fixed retroflexion of the uterus.

Ever since I have settled in Munich, I have observed an
extraordinary number of cases of isolated proctocolpocele. In
accounting for the genesis of this affection, we may unhesita-
tingly assume that it is produced by habitually protracted retention of the fæces in the ampulla of the rectum, the opening in
the levator ani being wide at the same time. Straining of the

abdominal wall during attacks of coughing, sneezing, etc.,
favor prolapse. A most pronounced case of this kind I observed
recently. Its interest lay in the fact that the uterus had been
drawn down by the proctocolpocele, precisely as it is otherwise
drawn forward by a cystocolpocele; and as in the latter case the
posterior part of the vagina remains almost in normal condition,
so here the anterior vaginal fornix and the anterior wall of the
vagina were approximately in their place. The opening in the
external os at the prolapse was in front and turned toward the
urethra, the uterus was greatly stretched, and the supravaginal
portion thinned. Here was at the same time retroflexion.

Concerning the relation of prolapse to retrodeviation, Fehling and Marion Sims regard retrodeviation as a consequence, or,
as it were, as a stage in the development of prolapse. Consequently, according to Sims, there is no prolapse without retrodeviation, while Fehling points out that the latter may be absent
in many cases.

The causal significance of retrodeviation which has been
set forth above all by Schultze and Kuestner and which is denied
by none of the other authors, may be easily understood from
what has been said hitherto. Displacement of the cervix in a
forward direction is naturally a common and frequent cause of
prolapse and retrodeviation, but here, too, it is difficult to explain
what, in retrodeviation, causes a prolapse of the uterus.

The perineal defect to which Duehrssen concedes so great
a significance, certainly does not give rise to it, because, as
already stated, it occurs so frequently without prolapse and in
prolapse may be frequently wanting. Utterly contrary to such
a view is the total perineal defect.

Schultze is of the opinion that in such a case parametritic
fixations prevent prolapse, while Kuestner thinks that the rigid

cicatrix of the margins of laceration hold the uterus in a posterior direction.

" Exact " is the favorite word of Schultze.   " I have tried ", he says, " to furnish an exact demonstration, and in the cases of total defect observed by me I never found parametritis and only once prolapse.   The latter is a very rare case; moreover, it must be stated that the total perineal defect is of thirty years' standing, but only a few years ago prolapse occurred, simultaneously with violent attacks of coughing during influenza. In most cases there existed nothing able to prevent prolapse; the margins of laceration were constantly soft, or at least never so rigid as to be able, like a pessary, to hold the uterus in an upward position.   In some cases there was even retrodeviation, a good chance for prolapse, and yet no prolapse actually took place."

The levator ani had remained intact and retained its functional ability; and although the laceration often extended up into the vaginal vault, the uterus remained in the pelvis.   In this case the cervix found a support in the posterior intestinal wall, instead of, as otherwise, in the posterior vaginal wall.

In conclusion, I shall make a few remarks on Schroeder's view concerning the origin of total prolapse.   My own experience, although comprising only two observations—such cases being exceedingly rare—teaches me that the uterus at the time of its deviation was larger and subsequently became very much involuted.   In the first case the patient was a lady 75 years of age, who had suffered from prolapse for at least half of her lifetime.   The trouble caused by this prolapse was aggravated by incontinentia alvi. The second case was represented by a young patient, in whom the first indications of prolapse appeared when she arose on the first day of her confinement.   In both cases the uterus was smaller than normal and in contrast with the

width of the aperture in the floor of the pelvic cavity. Likewise in all the other cases of total prolapse recorded and illustrated in medical literature, there is not a single one showing an
enlarged uterus. In the case described by Marion Sims, the
length of the uterus was only one and a half inch. Even Schultze's
cases confirm this fact, although he combats the theory. Further
proof for the correctness of our view may be found in the circumstance that enterocele occurs exclusively in cases of total
prolapse. The undiminished corpus with the annexa completely
and densely fills the aperture in the peritoneal infundibulum,
which latter may be seen from above. Nowhere is there any
space left through which a coil might pass, the corpus itself
being often kept back by the narrowness of the infundibulum.
Subsequently the corpus becomes smaller by puerperal or senile
involution while the aperture remains the same, and thus sufficient space is produced to allow not only the corpus but even
eventually intestined coils to slip through.

Finally with regard to the sharp distinction of the primary
sinking of the uterus from the secondary one as conditioned by
the colpoceles, which Fritsch points out so gravely, I am of the
opinion that, if the replaced uterus on pressure equally inverts
both vaginal vaults and the os uteri at the point of the prolapse
becomes deeper seated, this is no proof yet that such was likewise the case in the first development of the prolapse.

SYMPTOMS.

The symptoms differ very much according to the degree of
the prolapse and the manner of its origin. Investigating the
history of the case, the origin of prolapse may be traced back to
a confinement or to incipient senescence. However, great or
continual exertions of abdominal pressure, as coughing, vomiting,
lifting of heavy objects, may largely contribute to the production
of the pathological condition. But such excessive exertions, as

falling from considerable heights, lifting of a heavy weight, the
stopping of a rolling wagon, etc., are hardly ever the exclusive
factors of an acutely caused prolapse. The symptoms of the
latter are those of shock, vomiting, feeling of oppression, retarded
pulse, cold perspiration, etc. In the most pronounced degrees
of prolapse we frequently encounter only those complaints which
are caused by an unusual object lying before the genitals. Not
even vesical troubles are present, although the bladder is for the
most part hemmed in like an hour-glass and the urethra flexed
towards the lower part. Vesical catarrh may frequently set in,
and, as a consequence, pruritus, pain, and vesical tenesmus.
The latter may exist simultaneously with the inability of completely voiding the urine. Downward pressure, and a feeling as
if everything wanted to find an outlet below, so that the sick
dare not exert any abdominal pressure, as in cases of hernia, are
rarely wanting. On the other hand, the sensation of pressure
and pain in the region of the kidneys are of rare occurrence.

Of all the changes in position, prolapse is the only one that
may become fatal—and always by compression of the ureters.
The result is hydronephrosis, pyonephrosis, uræmia and death.
Kuestner observed three such cases.

If the annexa are diseased the complaints proceed from
them, and such complaints may be produced likewise without
prolapse.

The patients often state that before the appearance of prolapse they suffered from symptoms of parametritis, which disappeared later on. In such cases parametritis brought about cystocolpocele, the cystocolpocele led to prolapse, and the latter in
its turn again stretched the parametritic adhesions. But often
prolapse may occur right at the beginning, without any further
than mechanical molestation.

Those cases of cystocolpocele and parametritis constitute the complaints of parametritis and endometritis. (See above p.95)

PATHOLOGICAL ANATOMY.

Pathologic-anatomical investigations in cases of prolapse can be made much more rarely than in any other case of displacement, because a continued lying on the back almost always precedes death. In such a supine position gravitation ceases entirely and intra-abdominal pressure is diminished by almost two-thirds (in standing up it is 40 cm., in lying down 15 cm.), so that the difference in pressure between atmosphere and abdomen is scarcely perceptible. As a consequence, the outward lying organs are reduced in size, and a prolapse that could not be replaced previously may now be handled successfully by the physician, or be replaced by the patient herself, if it has not receded spontaneously. In the uterus thus restored to its normal position within the pelvis the characteristic changes are speedily involuted. For this reason I examined only two cases anatomically in which I supposed that prolapse had existed previously, because the uterus could be so easily drawn out (see Fig. 17 and compare Fig. 12 on plate VI in "Ueber normale und pathologische Anheftungen, etc.") and a Meyer's ring in the vagina sustained the supposition. In both cases the peritoneum over the bladder was not removable as usually, but adherent. Owing to this condition the bladder had formed a diverticulum-like pouch or excavation in the connective tissue of the ligamentum latum, so that under the ligamentum rotundum, under the tube and the ovarium as far as the Douglas (semi-lunar) fold, the peritoneum was undermined.

For a long time I examined every case of prolapse which I observed for the special purpose of ascertaining whether perhaps cystitis or local pelveo-peritonitis had led to the adhesion of the peritoneum over the bladder. If such had been the case,

the bladder, prevented from extending itself upwards, would
necessarily have extended downwards towards a place of less
resistance, which would have superinduced cystocolpocele and
prolapse. The formation of a diverticulum would thus have
been a consequence of the replacement by pessaries, the bladder
not being able to extend either upwards or downwards, and
therefore compelled to force a way laterally. Certainly in both
cases the process has not been otherwise. However, if I had
believed that a more general significance was to be attributed
to this etiological connection, such an expectation was not real-
ized. Some patients stated that a full bladder, others, on the
contrary, that an empty one caused a further protrusion of the
prolapse, while others maintained that they were not aware of
any influence exercised by that organ. I therefore soon abandoned all my investigations in that direction. Thus chance
furnished me, in a small amount of anatomical material, two
uncommon parallel cases. A third case of prolapse, the only
one I observed protruding outward during the lifetime of the
patient, and, after her demise, I examined in the summer of
1887 at the women's clinic in Munich. In this case I found the
bladder lying entirely outside of the pelvis. There was hypertrophy of the portio intermedia, extraordinary enlargement of
the ureters, and, if I err not, hydronephrosis. The patient died
of peritonitis in consequence of an attempt made to replace the
organ under the influence of anæsthetics, during which procedure old peritonitic adhesions were severed and an old ovarian
abscess on the left side was made to burst. Remarkable in
this case were the thickness and length of the muscular bundles
extending from the uterus to the peritoneum and the round
uterine ligaments. Unfortunately I have been unable to make
any observations since as to whether they always were in such
a condition.

TREATMENT.

Here, too, the modus operandi consists in massage and stretching of parametritis and irrigations for endometritis. Oophoritis and perioophoritis are treated by means of massage, and the ovary is loosened.

It is more difficult to find a correct prognosis in well pronounced prolapse. Can such a case be better treated by means of massage and stretching or by an operation? And in the great number of cases requiring operative treatment we may again ask what particular operation shall be performed?

The answer is very simple: In cases of anterior fixation of the cervix and retroflexion, massage constitutes the mode of treatment; in cases of too large an aperture in the floor of the pelvic cavity, operation. Between these two methods is another one, a bad but indispensable palliative procedure—the treatment by means of a pessary.

However, the application of this simple rule may be attended with difficulties, for on one side the fixations of the cervix in cases of outward lying uterus may again be stretched by dragging, while on the other hand the aperture in the floor of the pelvic cavity may be enlarged secondarily even in cases in which previously it had been narrow. One can hardly tell in single cases how prolapse occurred, and he must therefore pursue a tentative course. As already stated, in no branch of Brandt's mode of treatment occur so many failures; but on the other hand, there are cases in which the prolapse remains in the pelvis after the very first sitting. As at any rate, perhaps, existing concomitant diseases are removed, an operation should in no case be resorted to until the method of treatment has been applied for some time. Often hypertrophy of the portio vaginalis and intermedia recedes, and thus amputation may be dispensed with.

No doubt there are also cases in which both causes, displacement of the cervix and enlargement of the aperture in the floor of the pelvic cavity, were co-existent. Perhaps these are the most frequent. Above all, Brandt's method must be chosen for nulliparæ. I know such a case in which the operation was performed five times, and the anteriorly fixed cervix reappeared again and again, although hardly anything of the vagina is left. I also know a case in which the uterus and the greater part of the vagina had been extirpated, and yet the remainder prolapsed again. The poor patient! She was just about to be married.

BRANDT'S MODE OF TREATMENT.

His original directions are as follows:

1. Reach-support-standing, sacrum-beating.

2. Crook-half-lying, replacing of uterus in anteflexion, massage of the perhaps sensitive annexa and three to four times lifting of the uterus (Form I).

3. Crook-half-lying, knees-parting.

4. Bow-stoop-sitting, alternate twisting.

At home the patients have to make an injection of $\frac{1}{4}$ liter of cool water (about half a pint) in the morning and evening and perform the so-called pinchings (Germ. Kneifungen, Swed. Knippningarna).

This procedure differs from that employed for retrodeviation only in the manner in which the lifting of the uterus is concluded. In cases of cystocolpocele upward pressing of the plexus pudendus, and in cases of proctocolpocele S. Romanum lifting must be added as supplementary manipulations.

Light sacrum-beating, performed in convenient position, is said to have a tonic effect on the relaxed uterine ligaments. There being no such condition, sacrum-beating may be dispensed with.

Alternate twisting is a resistance-movement in the muscles

of the back. It is said to have a blood-diverting effect, and may, perhaps, by strengthening these muscles, bring about an improved posture. May be dispensed with, too, or be replaced by something less fatiguing.

Thus, besides the local treatment, there is nothing left except knee-parting and kneifungen.

Seated besides the patient that has already been treated by the manipulator, the latter, with his hands, draws her knees apart and asks her to offer as much resistance as is possible to her with free respiration. Thereupon she will bring her knees together, under the resistance of the physician. During this procedure she must elevate her buttock so high as to bring her shoulders, knees, and pelvis to the same plane if possible.

By this movement it is intended, by virtue of a kind of concomitant movement, to cause a contraction of the pelvic floor and thereby to strengthen it. It was this very movement against which Privy Councillor Olshausen rose in his might when Dr. Ahrens, before the Berlin Medical Society, reported upon it with exuberant enthusiasm. If, during the movement, we ascertain the condition by means of a finger inserted into the rectum or vagina, we shall find that only in a few rare cases is there a suggestion of contraction, and that mostly in cases of prolapse. In the great majority of cases the supposed effect is based on autosuggestion of the inventor. For the rest, it is a very good resistance-movement, in which almost all muscles of the back of our body are set in action.

However, it is different with the Kneifungen. They possess, moreover, the advantage that they may be executed by the patient without the aid of a physician, and they are of extraordinary value. They are performed in such a manner that the patient, lying with crossed feet on her back, executes the movements as energetically as she can, as if she wished to suppress the evacuation of the bladder and alvus, raising her buttock and compressing the nates, until at last she becomes fatigued. She then takes a rest, the movement having been made once. However, lest the patients become negligent, definite orders must be given. In the morning when the patient rises, and in the evening before she falls asleep, I cause the patient to execute the movement ten times each time of the day, subsequently even oftener.

By digital control-test during this movement we become aware how the sphincters contract and how the levator ani increases the flexion angle of the vagina and brings the apex of the angle nearer to the symphysis.

This movement proved quite successful in the case of a previously mentioned patient, 76 years of age and suffering from total prolapse and incontinentia alvi. Massage and stretching had been of no avail, and an operation was declined. Not only did a Meyer's ring remain in the vagina, but the incontinence disappeared entirely. In another case recently operated upon by me, a case of very rare occurrence, namely hypertrophy of the portio vaginalis, proctocolpocele threatened to set in after the removal of the portio vaginalis, which had caused a passive enlargement of the aperture in the pelvic floor. For some days "Kneifungen" were used, and the normal bending of the vagina was restored and thus prolapse prevented.

In a case of garrulitas vulvæ (vaginal flatus), without a perineal defect in the proper sense, but in which, through a number of rapidly succeeding births, the perineum had merely become relaxed, the vulva was silenced as it were by the "Kneifungen." These are only a few of the cases in which this simple movement contributed towards bringing about a cure.

Local treatment, however, constitutes the main thing. But even this includes yet some superfluous manipulations,

above all pressing of the plexus hypogastricus. In such a procedure both hands of the "movement-imparter" standing at the
side of the patient are, one above the other, placed on the abdomen of the patient lying on her back, and the tissues in the bottom of the pelvis are pressed somewhat in the direction of the
sacro-iliac articulation towards the pelvic wall. The supposed
effect is stimulation of the nerves for the purpose of strengthening the uterine ligaments.

Pressing of the plexus pudendus may be executed in two
ways. Either both thumbs are placed inwardly from the tubera
ischii, and the tissue pressed obliquely upwards and outwards
towards the tubera; or all four fingers of the half hand are distributed over the line between the two tubera, placed on the posterior perineum, and, as far as practicable, pushed up into the
pelvis.

Only to the latter kind of movement I ascribe any effect on
the pelvic floor and the articulation of the coccygeal bone—not
a nervous effect, but a mechanical one. For this reason it is
used by me.

Lifting has already been described on page 188. After it has
been performed three to four times, any pain caused by this
movement may be allayed by means of massage. This very
movement had to be modified, that is, simplified, because, performed according to Brand's directions, a considerable amount
of practice is required, and especially because it is executed by
two persons.

The modification given by Kuestner (1) I demonstrated as
early as 1887 in Munich, and it may be found also described
in Von Winckel's text-book. The assistant, standing beside
the patient and his face turned towards the pedal extremi-

_____

(1) Kuestner, Grundzuege der Gynaekologie, p. 91.

ties, executes the lifting with both hands, the margins of the thumbs touching each other.

Pawlick (2) executed the movement with a litte rod shaped like a drumstick, and Sielski (3) by means of a cup-shaped sound in order to prevent injury to the cavum uteri. E. Fraenkel (4) used bimanual lifting, the back of the outer hand turned towards the symphysis and the hands arranged in the manner in which the physician performing Brandt's lifting shows the uterus to his assistant, and lifting the organ higher up. I, myself, at that time, made use of Muzeux's forceps, applied to the portio vaginalis, for the purpose of lifting the uterus. The patient (5) did not understand how to relax the abdominal walls. I saw the patient again last summer. Only an insignificant cystocol-pocele, which had failed to yield to my treatment and raised doubts in me as to an ultimate and permanent cure, is left to-day, but of smaller size than at that time. Even to-day yet tests are made with the original lifting in my ambulatorium (clinic), if other methods fail; but so far I have met with no success in cases where manual stretching ultimately with use of forceps did not avail. In private practice, requiring an outlay not merely of time but of money also, I am not losing any time thereby, but I propose an operation, and only when the latter is declined I resort to the placing of a pessary.

Manual treatment, therefore, consists in bimanual massage of sensitive portions, stretching of palpable cords, pressing of the plexus pudendus, besides Kneifungen. Should nothing further be accomplished bimanually, irrigation treatment of the uterus is begun. The lengthened organ is thereby made to contract

(2) Pawlick, Centralblatt fuer Gynækologie, 1888, No. 18.
(3) Sielski, Centralblatt fuer Gynækologie, 1888, No. 14.
(4) E. Fraenkel, Ueber meine Behandlung des Scheidengebaermutter vorfalls, in Berliner Ærztliche Zeitschrift, 1888, No. 10.
(5) Volkmann, Klinische Vorträge, 353-354, p. 43.

and the rarely absent endometritis is removed. At the same time the cervix, seized with the Muzeux forceps, is vigorously pressed backward and upwards.

For the rest, Von Swicinsky[*] cured a prolapse by kneifungen and knee-partings, in short, by exclusive treatment of the pelvic floor. This is a further proof for the fact that there are cases where the size of the aperture in the pelvic floor, and again others where only displacement of the portio forwards may be the cause, although in most cases both causes are co-operating.

OPERATIONS FOR PROLAPSE.

Originally I gave the preference to Martins' operation because by it the columnæ rugarum are preserved and the excised strips of the vaginal membrane are taken from the lateral wall of the vagina. I intentionally placed the sutures in the region of the levator ani very deep, in order to include this muscle within the suture and to restore its tone.

The rectum in the region of the vaginal fornix adheres firmly to the posterior pelvic wall. This portion of the rectum in case of a prolapse does not follow the posterior vaginal wall. Proctocolpocele has a different genesis. Moreover, it is necessary that the connection through short and solid muscular bundles, which, in an uninterrupted succession from the levator ani to the vaginal fornix, joins the vagina and the rectum be severed, so that the vagina may formally detach itself from it, as may be seen in the rarely occurring cases of total inversion of the vagina without proctocele after replacement (Fritsch). To me it seems to be more likely that this detachment occurs likewise under the influence of senile atrophy of those muscular bundles. In every prolapse the posterior fornix, for a few centimeters, descends deeper. This process only needs continuing

*Von Swicinsky, Muenchener mediz. Wochenschrift, 1889.

downward. However, there are total prolapses without total inversion of the vagina, and total inversion of the vagina in incomplete prolapsus uteri, as Schultze has found out and demonstrated by cases.

Although I could not exactly complain of failures in the application of this method, the deviation of the vagina backwards after the operation being every time very plainly re-established, I nevertheless abandoned this sort of treatment. Some cases of proctocolpocele that had come under my observation in brief succession, induced me to take this step. I must confess that here I had no confidence in the S. Romanum liftings. In the first cases I felt called upon to operate, especially as I was confronted with a partial perineal defect which could not have been removed by manual therapy. During the operation itself I endeavored to form a pretty thick pad, a living tampon, as it were, which was to push the extended rectum backwards. Thus I came to the operation which Frank had described in 1887, during the meeting of naturalists at Wiesbaden. An arched incision along the posterior margin of the vaginal entrance and the drawing apart of the wound margins give a good start for the perineum to be formed subsequently, and enable the advancing finger to find easily the scant layer of long-fibered connective tissue which lies between the rectum and vagina. Bluntly and boldly advancing, the exploring finger easily detaches the vaginal tube from the intestinal tube, at least up to the proximity of the portio vaginalis or at least up to and beyond the proctocele. It is easy now for the two index fingers to enter, and, by lateral pressure, to enlarge the funnel.

The operator beginning from above, now unites the continuous lateral walls by deep catgut sutures, the stitches being each time knotted by looping of the thread. I was astonished to see

how easily the margins of the lavator ani could be found by
means of this method. By a few far-reaching stitches I seized
them and brought them up into the suture. At last there
remained a crescent-shaped wound area, the horns of which
crescent lay on the labia minora. This wound area, by means of
horizontally executed sutures underneath it, is so stitched
together that the horns are made to lie upon each other. In
this way a broad, solid perineum is established.

The good results thus attained induced me to proceed in the
same way in each case of prolapse, and to apply the same
method also on the anterior wall of the vagina, for cystocolpocele; anteriorly, however, without the formation of a perineum-
like eminence, which may easily be accomplished by appropriately placing the last sutures closing up the vaginal wound.
The two pads or cushions thus created, exceedingly strong columnæ rugarum as it were, press the cervix vigorously backward
and thus form a base, a living pessary, such as Simon, Hegar
and others wished to attain by their methods.

Fritsch, too, I am pleased to say, in the latest edition of his
textbook, recommends the operation of Frank. The suturing of
the margins of the levator ani is likewise recommended, although
otherwise Fritsch does not anywhere point out the importance
of this muscle for the remaining of the uterus within the pelvis.

Only I am astonished to see that Fritsch, in case of proctocolpocele, resects a portion of the vaginal cushion. The latter, owing to its thickness, afforded me the very best means to
press the rectum backwards, and I never was obliged to narrow
the latter by superficial horizontal sutures.

Disadvantages such as lung embolisms, mentioned by
Bumm,* I never observed The operation is less complicated and
less bloody than any other one. I carefully endeavored to avoid

---

*Centralblatt fuer Gynækologie, 1894, No. 29.
14

the formation of dead spaces, taking up with the needle also some tissue from the anterior and posterior walls of the funnel.

I shall only briefly mention the other operations. Hegar cuts a triangle from the posterior vaginal wall, the apex near the portio and the base along the posterior circumference of the ostium vaginæ. Simon cut a triangle obliquely truncated at the top so as to produce a pentagon, with the base lying along the hymenal remnants. Fritsch added to it a powerful plastic perineal operation, giving rise to a figure which resembles the sagittal section of a bell. However, an older method of Graily Hewitt insured almost the same incision. The wound margins to the right and the left are sewed together by vaginal and perineal sutures, so that the figure in the middle is, as it were, folded together.

Other authors preserved the columna rugarum posterior, as Graily Hewitt by means of a second method, Bischoff, and A. Martin. Hewitt and Bischoff cut triangles, Martin parallelograms from the lateral wall of the vagina. The methods of Hewitt and Bischoff differ from each other only in so far as in the latter's operation the perineal incision does not extend forward so much. In cases of insufficient narrowing of the vagina fusiform flaps from the posterior wall are excised and the wound margins united.

An older method of Von Winckel's was, properly speaking, not much more than an episiorraphy. The first cut runs along the posterior circumference of the introitus, a second one parallel four cm. higher up. A sagittal cut unites both in the middle. From this sagittal incision the flaps thus cut around are dissected back, shortened, the wound surfaces from both sides joined and sutured together, and the remainder of the flaps tucked up and sewed together so as to give rise to a small rump.

A second operation consists in cutting out, in the middle of the vagina to the right and to the left, a semicircle, about three cm. wide, from the vaginal tube. Only in the middle a piece of the vagina is left, both anteriorly and posteriorly. The wounds of the anterior and posterior walls are now sewed one upon another, producing stenosis of the vagina. If the latter should happen to lie within the confines of the levator, the operation may be called a good one theoretically.

With regard to the latest methods the following may be stated. The vaginal portions cut around in the operation of Hegar and others are not detached, but merely dissected back. The wound margins thus created are, as in the preceding case, sewed together and the circumcised vaginal membrane thereby imbedded. This modus operandi, however, is not to be recommended, as it artificially produces dermoid cysts.

Even Marion Sims, in his mode of operation, allowed the center of the incision made in the shape of a mason's trowel to stand and merely sewed together the small wounded margin, but towards the uterine orifice he left an opening through which the "secretion" might flow off. Here, too, we find the idea to utilize the resistant doubled-up vaginal membrane as a living pessary, as it were.

An operation calculated simply to narrow the vagina promises from the very beginning but little success, if we consider the degree of expansion of which it is susceptible by births and tumors. If narrowed, it may be easily enlarged again. The enlargement of the vagina is not the cause, but the consequence of prolapse.

In the operations of Spiegelberg, Neugebauer senand Le Fort corresponding surfaces of the anterior and posterior vaginal walls are incised and sewed up together. A bridge of flesh is thus established about in the middle of the vagina. Such may be called palliative operations.

The Alexander-Adams operation and the ventrofixations afforded no relief for prolapse. The uterus extends and partly prolapses, in spite of the fundus having been fastened anteriorly. If we are obliged, besides, to call colporhraphy to our assistance, this constitutes no success, the latter without the remedying in most cases former operation. The measures taken against a non-fatal or a rarely fatal complaint must be free from danger.

Total extirpation I have mentioned previously.

PESSARY TREATMENT.

If manual treatment and operation are declined, or if the former is a failure and the latter declined, a pessary must be inserted. If, after replacement of the prolapse, retrodeviation takes place, the latter, too, must be replaced and such a pessary be inserted as will compel the uterus to keep in anteflexion. (See p. 199.) Here, even *after* the climacteric period has set in, retroflexion must be corrected, whereas in cases of retroflexion without prolapse in menopause it does not matter how the uterus lies.

If the uterus lies anteriorly or if such a pessary is not tolerated, a Meyer's ring is inserted. If the latter should prove unbearable or drop out, a tampon is introduced, changed every two days, impregnated with acetate of aluminum, and eventually fastened with a T-bandage.

APPENDIX.

Brandt's Treatment of Incontinence of Urine, of Floating Kidney and Hernia.

Incontinence of urine is quite common also in cases of chronic parametritis and other inflammations in the pelvis. Vesical tenesmus, so frequently mentioned in this book, may become so intense as to lead to involuntary urinary discharges. In such cases bimanual removal of the inflammation suffices in order to bring about a simultaneous discontinuance of incontinence.

In uncomplicated cases Brandt recommends bending the index finger of the left hand from the vagina around the symphysis in such a manner that the finger tip may reach its superior margin. Groping gently around one feels now the neck of the bladder and the urethra slipping through under his finger. With firm pressure of the finger tip the neck of the bladder is now pressed several times toward the bone. If the right hand encircles the wrist of the left hand, pressure is exercised with a greater degree of certainty. This pressure of the sphincter vesicæ is accompanied by blood-conducting movements. I attained surprising cures by a single pressure without hygienic gymnastics, and experienced failures in spite of blood-supplying movements. For this reason I do not employ them.

In two cases I successfully used Sænger's treatment—stretching the sphincter in all four directions with inserted catheter; so vigorously, indeed, that the urine ran out beside the instrument. True, in one of these cases bleeding was caused. In one case I placed a catgut around the region of the sphincter and kneaded (massaged) over the sound; in another case I used fil de Florence (silk-worm gut). In the latter case no cure was effected and nothing was left but to try Von Winckel's, respectively Schultze's, treatment. It consists in narrowing the urethra by excising a piece of the wall and sewing up the defect. The last case I treated by pressure was that of a girl fifteen years of age, who was afflicted also with nocturnal enuresis. The very first pressure from the rectum effected a cure of both complaints. This mode of treatment was applied only three times altogether. Brandt recommends the latter also for nocturnal enuresis of children.

In cases of *floating kidneys* Brandt has suggested a diagnostic and a therapeutic manipulation. For diagnostic purposes the loins must be clasped in such a manner that the four fingers lie between the pelvic bone and the false ribs and the thumb below the hepatic margin previously determined upon. The patient, with drawn-up knees, occupies a lithotomy position. She is asked to breathe in the air deeply. At the height of the inspiration the thumb enters obliquely backwards, and upwards under the margin of the liver, the abdominal cavity, the kidney, if movable being now for the most part pushed a little towards the middle below the thumb. Here it may be touched by the gently groping right hand and its form determined. If we raise the thumb slightly, the kidney will glide upwards. This very circumstance, the ability of its being replaced, is the characteristic feature, very slight degrees of movability being often found in the thickest of abdominal walls.

This manipulation enables us to draw a conclusion with regard to the obscure etiology of the displacement. The fact that running, exertion of abdominal pressure, continued standing up, etc., in the male sex operate in the same way, but displacement is never met with; and the further fact that displacement may also occur in nulliparæ, where pregnancy and pendulous abdomen cannot have any influence, invest the above named injurious effects at most only with very slight significance. But the circumstance that displacement occurs as a result of compression of the hepatic region under the action of the diaphragm; and that, moreover, it is found only in the right side, justifies us to draw the conclusion that the liver and the diaphragm, as affected by *lacing*, displace the kidney.

In my course of examination this manipulation is constantly studied and practiced, and every woman examined for movable kidneys. The latter may frequently be met with where no complaints of any kind pointed to it. In other cases the patient complained of pains, while walking in that particular region, and of a sensation resembling that produced by a dropping body; often also of pains under the shoulder blade, as in the fourth and fifth intercostal spaces, anteriorly. The latter symptom especially interested me very much, because I learned once from a patient that Professor *Seydel* in Jena had, from this complaint alone, without making any examination, diagnosed floating kidney in her. I know Professor Seydel as a very cautious diagnostician. Thus in no case in which the above pains are complained of did I fail to find floating kidneys. However, this does not signify by any means that the above described pain is a necessary concomitant of every floating kidney, cases without any pain at all being frequently encountered.

Performing the therapeutic manipulation, the physician, standing to the right of the patient, places both his hands, one over the other and the finger tips pointing headward, on the abdomen. Under oscillating movements (in order that the intestinal coils may get out of the way) they enter beside the vertebral column and the finger tips are placed *below* the kidney into the channel beside the vertebral column, and thus the kidney, under vibratory pressure, pushed upwards under the liver as far as possible.

In fresh cases the displacement is cured; in older cases only the complaints disappear, but always and after a very short time.

This mode of treatment may be appropriately learned and applied by the patients themselves.

As additional treatment Brandt uses hygenic-gymnastic movements of the rectus and obliquus muscles of the abdomen. The patient sits astride on a bench. Her knees are fixed by a second person standing before her, or stirrups are fastened on the floor into which the feet are inserted. The patient having bent herself forward, the attendant behind her seizes her shoulders and draws them backwards, the patient resisting. Under the resistance of the gymnast she then bends herself again forward.

The oblique abdominal muscles are exercised in such a manner that only the shoulder of that particular side is bent forward and drawn back under resistance.

This movement, too, may be executed by the patients alone. For this purpose they lie down on the back, and without taking hold of a supporting point or crossing their arms on the chest, alternately raise and drop the upper part in slow succession of the body. Respiration must not be checked. The oblique muscles are exercised, the shoulders of that particular side being bent forward and drawn back.

I describe this movement so accurately because it renders excellent service for the involution of the pendulous abdomen after childbed. I give here a definite direction—to execute the movement in the morning and in the evening while lying in bed, keeping ten times head and upper part of the body straight and ten times with shoulders bent forward.

In the latter case intestinal massage may be added.

In cases of hernia Brandt first replaces the rupture. Here, too, in difficult cases, he has given good directions for taxis. The patient lying on her back and raising her buttock considerably, a movement is executed above the rupture which is quite analogous to the S. Romanum lifting. Taxis is followed by vigorous massage of the hernial ring and surroundings, whereupon the above described hygienic movements are executed, for the purpose of strengthening the muscles and pillars of the hernial ring, thereby narrowing the latter.

INTESTINAL MASSAGE.

For Constipation Fulling, Stroking of the Colon, and Massage of the Small Intestine are Employed.

Fulling is done in this manner: The kneading (massaging) person, sitting or standing on the left side (of the patient ready to undergo the treatment), places both hands, one over the other and at the same time one a little before the other, transversely on the abdomen in the umbilical region. Alternately pressing his finger tips on the right side, he endeavors to draw the whole contents of the abdomen to the left, and thereupon pressing with the balls of his hands, he seeks to shove everything back to the right side. Is continued for from two to five minutes.

Colon-stroking begins first with stroking of the colon descendens, followed by repeated stroking of the transversum and descendens, and finally by that of the ascendens, transversum, and descendens. *No* interruption in the pressure at any place must be allowed, because otherwise the intestinal contents might slide back. Is executed ten times in succession. Only in this movement the hand is moved against the skin of the patient.

Massage of the small bowel, perhaps the most dispensable of the three, is performed by placing both hands crosswise in the umbilical region. The margin of the hand is impressed, continually following a circular course, as if one wished to impress his hands along the margin of a small disk. Is applied for four to five minutes.

All the last described treatments are such as may be easily learned and employed by lay people or by the patients themselves.

TABLE OF CONTENTS.

— — —

INDEX.

www.ingramcontent.com/pod-product-compliance
Lightning Source LLC
Chambersburg PA
CBHW030354270326
41926CB00009B/1099